GATLING

A PHOTOGRAPHIC REMEMBRANCE

E. Frank Stephenson, Jr.

Meherrin River Press

Book Layout and Design by
E. Frank Stephenson, Jr.

Published by
Meherrin River Press
301 East Broad Street
Murfreesboro, North Carolina 27855
(919) 398-3554

Library of Congress Number 93-79035
ISBN 0-9637671-1-9

Printed in the United States of America

Pierce Printing Company
Ahoskie, North Carolina
1993

PREFACE

Few native sons of North Carolina have ever achieved the fame, fortune and international recognition of Richard Jordan Gatling, often called the Robert Oppenheimer of the nineteenth century. Born 1818 in a log cabin in Maney's Neck Township (Como) in the northern section of Hertford County, Richard Jordan Gatling mounted a steed in 1844 and headed west to St. Louis, Missouri. There he made his first fortune in seed sowers, mainly wheat drills fashioned from a cotton seed sower he had developed and patented back in Hertford County. His successful quest for a new life was unprecedented, as Gatling, in the last half of the nineteenth century, was recognized as one of the world's greatest inventors. Combining the principles of his father's rotary cultivator and his own seed sower, Gatling's invention of the equivalent of the atomic bomb of his time, the Gatling gun, in Indianapolis, Indiana in 1861 and 1862 forever changed the way men of all nations waged war. But what was supposed to have made war obsolete, only made it more destructive. This however did not lessen the fame and fortune of the man from the rich, peanut producing, farmlands of the Meherrin River region of North Carolina as Gatling's name became a worldwide household word from the courts of kings, queens and czars to the humblest of homes in America. This is contrary to what some of Jemima Sanders' Indianapolis friends told her on the eve of her marriage to Gatling on October 24, 1854, "that Gatling nor any of his strange inventions would ever amount to anything." Today the name of Richard Jordan Gatling has all but disappeared into relative obscurity along with the gun's nicknames of Gat, cornsheller and coffee grinder. In 1911 some fifty years after its invention, the Gatling gun was declared obsolete by the United States Army and the gun like its inventor soon became long forgotten. It would take almost another fifty years for the Gatling gun to emerge from the obscurity that began in 1911. Gatling's name, not his gun, enjoyed a brief reprieve during World War II when the federal government named two ships for him, a destroyer and a Liberty ship. However it was the extensive rapid-fire armament research following World War II that returned the Gatling gun to prominence. Its modern form, the Vulcan gun, is now standard equipment in the armed forces of the United States. An odd footnote in the Gatling gun saga occured in October 1950. When the Chinese Communists swarmed across the Yalu River to enter the Korean War they brought with them some vintage Gatling guns that Richard Jordan Gatling had sold in China in the 1870s. This photographic remembrance is presented as a tribute to Richard Jordan Gatling, an inventive but humble man who was clearly way ahead of his time, and to his remarkable family whose attributes of honesty, perseverance, kindness, hard work, dedication and compassion live on today in the Maney's Neck Township of Hertford County, North Carolina.

IN MEMORY

ELEANOR HULINGS GATLING
June 1, 1903, Pittsburg, Pennsylvania
May 2, 1993, Asheville, North Carolina

For fifty-nine years she proudly shared the Gatling name as the wife of John Waters Gatling, grandson of Richard Jordan Gatling. John Waters Gatling and Eleanor Hulings were married April 19, 1934 in Philadelphia, Pennsylvania. Photograph taken at Federal Shipbuilding and Dry Dock Company, Kearny, New Jersey on June 20, 1943, shortly after she had christened the USS Gatling (DD671) a destroyer named in honor of Richard Jordan Gatling. Burial was in Crown Hill Cemetery, Indianapolis, Indiana. Photograph courtesy of John Waters Gatling.

DEDICATION

This publication is dedicated to the memory of two men whose Gatling stories stirred the mind of an inquisitive North Carolina country boy.

F. Roy Johnson
Publisher-Printer
Murfreesboro, North Carolina
November 12, 1911 - October 17, 1988

E. Frank Stephenson, Sr.
Farmer-Craftsman-Cook
Como, North Carolina
October 26, 1909 - April 13, 1990

APPRECIATION

I would like to express my profound appreciation and sincere gratitude to Richard Jordan Gatling's grandson, John Waters Gatling whose generous sharing of family information and Gatling archival material made this book possible.

Grateful appreciation is also extended to the following for their invaluable assistance:

Dr. Thomas C. Parramore, Raleigh, North Carolina
Mr. Robert Lee Gatling, Ahoskie, North Carolina
Mr. Colbert P. Howell, Raleigh, North Carolina
Mr. Harold Gatling, Murfreesboro, North Carolina
Mrs. Elsie Taylor Parker, Murfreesboro, North Carolina
C. B. Robertson Family, Jackson, North Carolina
Mr. Wayne Sanford, Indianapolis, Indiana
Mrs. Julie West, Murfreesboro, North Carolina
Ms. Marion Chertow, Hartford, Connecticut
Mr. Matthew Memerson, Hartford, Connecticut
Mr. Dennis Main, Hartford, Connecticut
Mrs. Renee Main, Hartford, Connecticut
Col. James C. Pennington, Murfreesboro, North Carolina

E. Frank Stephenson, Jr.
August 1, 1993

We Grew Up In The Same Neck Of The Woods

E. Frank Stephenson, Jr.

1993

I probably know the old Jordan Gatling plantation in Maney's Neck Township, Hertford County, North Carolina where Richard Jordan Gatling was born and raised about as well as anyone in the United States today, and I have been on it at least a thousand times during my lifetime. My association with and knowledge of the Jordan Gatling plantation stems from the fact that I literally grew up on a peanut and tobacco farm next door to it in the 1950s, and, during my teen years, my family farmed it. Having grown up next door to the old Gatling plantation, it always seemed like a trip back home each time I would visit it, and today's visit to it is no exception save for one very sad difference. As I slowly walked along the old dusty dirt path leading past the Gatling family graveyard I angrily stared ahead at the empty spot where the old Jordan Gatling house had stood until 1979. I momentarily stopped walking as my thoughts gradually drifted back to my earliest recollections of this once proud plantation.

As I wiped the sweat from my face in the torrid July heat, I glanced back at the old Gatling family graveyard and wondered if all of the local tales and ghost stories about the Gatlings I had heard while growing up were actually true. I chuckled as I remembered hearing my Grandmother Stephenson telling the story about James Henry Gatling, Richard's brother, building a horse drawn four-wheel farm wagon in the cellar under the Gatling house and when he had completed it, he could not get the wagon out because it was too big to go through the cellar door. I briefly recalled another story Grandmother Stephenson had told

while I was a youngster. This story was about the time when one of Richard's deceased sisters was being buried in the old family graveyard. Supposedly it took twelve slaves to carry her jewel-filled coffin from the house to the cemetery. I never really believed that Gatling tale was true, but the strange thing about it is that forty years ago the same sister's coffin was dug up, robbed, and left standing upright on its top end. Some of the neighbors quietly reburied the coffin. I recalled the slave stories, the folklore and the multitude of ghost stories particularly centering around Richard's murdered brother, James Henry, and how these had dominated the local homespun conversations for years and years creating a cult-like atmosphere among the followers of this creative but eccentric family.

Among the local black and white populations of Maney's Neck Township, there exists today a treasure of Gatling ghost stories as many people believe the old Jordan Gatling plantation and particularly the old family graveyard are haunted with hants. James Henry Gatling's violent death in 1879 has spawned a repertoire of ghost stories especially about, but not limited to, him. Supposedly a slain and bloodied James Henry Gatling on horseback can be seen riding at night along "Gatling Avenue," the one-mile long raceway that the Gatlings built to race their fine stock of race horses. Those who have witnessed this event declared that the horseman was actually James Henry Gatling's ghost with an axe stuck in the top of its head. Around 1900, some years after James Henry Gatling was murdered, a black coon hunter and his old coon dog were hunting one night on the old Gatling plantation. The old black man had known James Henry Gatling before he was murdered. Suddenly the coon dog picked up a trail deep in the woods. The trail weaved its way through the woods to the Gatling cemetery. Reaching the old graveyard, the coon dog began barking up one of the huge old magnolia trees in the old graveyard. The coon hunter positioned his lantern to get a fix on the treed animal's eyes but he could not locate any eyes among the heavy, thick green magnolia leaves. Determined to find out what his dog had treed, and with a fairly bright moonlight, the coon hunter set his lantern down on the ground and proceeded to climb the tall magnolia tree in search of a coon or 'possum which he knew just had to be up the tree because his old coon hound was among the best coon hunting dogs in this part of the state. Slowly but surely the crafty old coon hunter climbed up the tall tree carefully checking each limb as he went but he saw nothing. Just as he was about to climb down he heard a voice from above that sounded just like the late James Henry Gatling, "Uncle, come up a little higher, I'm up here!" The shocked old coon hunter fell to the ground and outran his coon dog home. Folks around the neighborhood reported that the old coon dog never did hunt any more. They said it probably was caused by the fact that the old coon hunter never did either.

Although the ghost of the slain James Henry Gatling seems to dominate the Gatling ghost stories, there are many other Gatling ghost stories centered around the old Jordan Gatling plantation. There is constant talk among the locals about "hants a-walkin at night" up and down the long dark path leading to the Great House. Some believe these ghosts to be those of the Gatling slaves who were buried on the outside of the Gatling family cemetery.

The old Jordan Gatling house itself was thought to have been haunted. This writer, who hoed corn, cotton and peanuts on the old plantation, heard, as did others, unexplained noises coming from the vacant old house. These noises included the slamming of doors in the vacant house when there was no breeze blowing at all. Other bumps, thumps, screams, yells, moans, groans, shrieks and wierd laughter could be heard coming from the old house. The old barns and outbuildings which were located on the plantation also were haunted. One Gatling ghost story tells of an old black man shucking corn on a cold winter morning in a log barn that stood near the Great House. Suddenly the old man bolted from the barn yelling and screaming and all covered with broken glass. He explained that a ghost had "busted" an empty fruit jar over his head. Even the woods surrounding the old plantation are reported to be haunted. It seems that in the 1940s and 1950s when illegal bootleg whiskey operations were at their peak in Hertford County, the bootleggers who had their "stills" hidden deep in the Gatling woods were constantly encountering unexplained noises, lights and movements. One such bootlegger had to move his still from the Gatling woods because his still operator, a black man, refused to go in the Gatling woods any more. The bootlegger asked him why he did not wish to go back in the Gatling woods. The still operator told him that he had seen "hants" drinking from his mash vats on more than one occasion and that one of them, a woman, had offered him a drink from an old gourd dipper. The bootlegger accused his still operator of "sampling too much of his own stump juice" but he never did persuade him to return to running a still in the Gatling woods. And today among the local deerhunters who occupy "deer-stands" during hunting season along Gatling Avenue swear that they can hear the pounding of horses' hooves along the old raceway.

How much credence can be placed in the Gatling ghost stories is open to speculation. But one thing is certain, the Gatling ghost stories remain as popular today among the old local story tellers as they were fifty years ago.

As I stood there in the rutted and well-worn path, my thoughts went back to a time when some of my black playmates and their parents occupied the old house; when the soggy fields were tilled; when catfish and perch could still be caught in the two old fish ponds nestled deep in the yellow pine woods surrounding the open farm lands; when remnants of the magnificent trees planted by Jordan Gatling were still lining the original lane leading to the house from the Murfreesboro, North Carolina-Petersburg, Virginia road; and when other buildings were still standing on the place. It was also a time when I had worked the fields on the old plantation; I had fished in the old fishponds; I had found a handful of Confederate money that James Henry Gatling had secretly stashed in a niche in the floor joists to prevent "Buffalo" raiders from stealing it during the Civil War; I had fished and swam in the same fishing and swimming holes along the beautiful but wild Meherrin River where Richard Jordan Gatling had fished, swam and perfected the screw propeller a hundred and thirty years earlier; I had begun collecting Gatling stories and memorabilia; I had heard all of my life the Gatling folklore and ghost tales; I had purposely visited the old plantation for over twenty-five consecutive Christmas holiday seasons; I had made a gallant effort to purchase the old Gatling house to restore for my own home and, yes, I had shed a tear or two the day the very same old house was pushed over and set ablaze. Although I had never lived in the old Jordan Gatling house, it was and still remains, as much a part of my life as the house where I grew up just a few hundred yards west. I absolutely will never forget as long as I live the brutal destruction of this unique plantation and its finely crafted and sturdily-built two-story frame house where one of the world's greatest inventors, Dr. Richard Jordan Gatling, grew up.

To reach the old Gatling plantation, one travels on U.S. 258 four miles north of Murfreesboro, North Carolina through the rich and lush peanut-producing farmland of Maney's Neck Township. In the days when Richard Jordan Gatling was growing up on the plantation, Murfreesboro, an 18th century maritime village on the Meherrin River, was the center of commerce, education, culture and transportation in the region. Today, Murfreesboro, the home of Chowan College, is primarily a college town with some light industry and a successful preservation program to save its early structures, many of which predate or date from the days when a young Richard Jordan Gatling was a frequent visitor to the town and was a law clerk there in the office of Lewis M. Cowper, his great uncle.

Surprisingly, Richard Jordan Gatling still has a number of kinsmen residing in North Carolina including his grandson, John Waters Gatling in Asheville, and a hardy sprinkling of cousins, particularly in his home county and the neighboring county, Northampton. Today, the late Dr. Richard Jordan Gatling, in his native county of Hertford, is affectionately recalled as "Cousin Dick Gatling" by his Tar Heel relatives. Numerous people of the Gatling name in the United States often claim direct kinship with the world famous inventor. Obviously, every person in the United States bearing the name of Gatling is not a descendant or a direct relative of Dr. Gatling. Thus, one of the purposes of this sketch is to provide a brief historical sketch of Dr. Gatling's North Carolina family and relatives beginning with his grandfather, James Gatling.

As far as can be determined, Richard Jordan Gatling's grandfather, James Gatling moved from Virginia into the Maney's Neck or the northern part of Hertford County a few years before the American Revolution. While his exact birthdate is unknown, James Gatling married Mary Cowper of Maney's Neck Township in 1778. Five of their children reached adulthood, including Elizabeth, Polly, James, William, and Jordan, who was the father of Richard Jordan Gatling. An industrious and frugal man, Grandfather James Gatling acquired, through hard work and self-sufficiency, a modest holding of land. Much of his land was in forest and needed clearing to make it productive. Before his death in 1822, James Gatling and his family were living in a comfortable clapboard house a mile south of Buckhorn Chapel, now known as the village of Como in Hertford County. His home was a tall roomy structure with a "jump" or an upstairs area commonly used for spare sleeping quarters. The house stood until the early 1950s.

In a list of taxable property for Hertford County in 1782, the first year after the Revolutionary War, it shows James Gatling to have been in possession of about three hundred acres of land. He and his sons must have labored hard alongside the one slave man of prime working age who he had listed. James Gatling was one of six Gatling families residing in Hertford County with one slave man each. He was slave poor as the average slave holding for District 1 where he resided was 3.48 per list. Yet, only twelve landholders of District 1 had ten or more slaves and only two of these owned more than twenty. They were, James Maney with 2,531 acres and twenty-nine slaves and George Little with 2,291 acres and twenty-two slaves. James Gatling rated quite well with working stock-animals. He owned three horses and mules while the average for the district was 2.65. He had five head of cattle while the average was 8.5. The value of his personal property was $405.00 while the average was $374.50.

When the first United States census was taken in 1790, James Gatling's household consisted of one free white male over sixteen years, three white males under sixteen, three white females, and two Negro slaves, perhaps a couple. Thirty-two years later James Gatling had no less than five slaves. His will, dated February 12, 1822 states "it is my desire that the division of my Negroes be equally among my five children should be without sale." Records further indicate that none of James' sons, Jordan, William or James, nor any of their children, ever sold their slaves outside of the Gatling family.

When James' son, Jordan, the father of Dr. Richard Jordan Gatling, died in 1848 at the age of sixty-five, he had by natural increase or possibly purchase some twenty slaves. He was listing nine black males and the remaining were women and children. Several years before emancipation Jordan's son, James Henry, willed that none of his Negroes should be sold. After the Civil War, Richard Jordan Gatling was still remembering his late father's former slaves with gifts of clothing, housewares and health supplies. The Gatling family's former slaves were not the only persons in Hertford County that Richard Jordan Gatling, while residing in Hartford, Connecticut, was remembering with gifts. Quite often he shipped trunks full of clothing, household goods and other materials to his North Carolina cousins. Several of these trunks survive today.

Richard Jordan Gatling and his brothers' loyalty toward the family's former slaves seemed to have provided some stability among the black Gatlings. They possessed a sense of belonging to the white Gatling family. This feeling was very evident more than a century after Grandfather James' death in 1822. In May, 1932, Uncle Tom Gatling proclaimed himself as the "born guardian of the Gatling cemetery" at the old Jordan Gatling homestead in Maney's Neck Township. Upon this occasion Uncle Tom Gatling told a young visitor, Richard McGlohon of Winton, "Naw suh, I hain't gwine let nobody tamper wid de things in my grabe yard. Dem's my folks and I'se sweared to look atter dem." The Gatling slaves were buried on the outside of the elaborate eight-sided iron fence enclosing the graves of Dr. Gatling's parents and all of his brothers and sisters, except William, who is buried in Canada. Some Gatling slave descendants today live just a stone's throw from the old Jordan Gatling plantation.

Perhaps persuaded by the Great Revival which swept across Hertford and various other North Carolina counties early in the Nineteenth century, Grandfather James Gatling was a deeply religious and God-fearing man. He states in his will of 1822, "First of all I do recommend my soul to God who gave it in hopes of receiving a joyful Resurrection with the redeemed in Christ."

Grandfather James Gatling died in 1822 and was buried on his homestead. His grave is marked by a concrete slab placed there around 1900. Of the five of Grandfather James' children who reached adulthood, very little is known about Elizabeth, William, and Polly, who married Charles Gay. James, the oldest son of Grandfather James, inherited his father's homestead, married, and settled there. When he died in 1855 and was buried next to his father, he had left his father's old homestead to his only son, Roschus B. Gatling who was Dr. Gatling's second cousin. Roschus B. Gatling, who was a Hertford County constable, had eight children and many of their descendants today have archival materials and artifacts relating to Dr. Gatling and his brother, James Henry Gatling.

When Grandfather James Gatling's other son, Jordan acquired his first plantation, he would have preferred the well drained and cleared lands along the banks of the Meherrin and Chowan Rivers or some major creek tributaries. But most of this naturally or easily drained land was either too expensive for him or owned by large slaveholders such as the Maneys, Murfrees, and Littles. Much of this same land had been taken from the Meherrin Indians about three quarters of a century earlier. In 1724 these Indians were complaining, "their Lands were all taken up and Surveyed by the English and that they were forewarned from off their clear Grounds and forbid to plant corn thereon."

Thus, Father Jordan Gatling's plight was like that of many other small farmers; he had to content himself with eighty acres of soggy and undeveloped flatlands. Their value on the tax books was about fifty cents an acre. This damp and heavy soil was unsuited for agriculture until a network of drainage ditches and canals had been cut and the forest cleared. The original eighty acres that Jordan Gatling purchased was known as the old Crutchlow tract. Jordan Gatling delayed marriage several years, and, with extraordinary energy and perseverance, he felled the trees, cut the ditches and toiled with his yoke of oxen until he had carved out a plantation.

While clearing the land for his fields, Jordan

selected prime yellow pine trees to use in building a 16 foot by 20 foot log cabin which he covered with slabs from rot-resistant cypress trees. Although in later years Jordan was regarded as somewhat unsociable and odd, his neighbors may have kept to the custom of time and helped him with his house raising. This was the log cabin to which twenty-eight year old Jordan would bring his bride of fifteen, Mary Barnes, in 1810. This cabin was similar to the log cabins of the frontiersmen a century earlier. It was a crude but comfortable shelter with a large mud and stick chimney to serve for heat and cooking. The exact location of Jordan Gatling's log cabin is known by this writer.

As a youth under the tutelage of his father, Jordan Gatling had learned that the small planter survived and prospered by the measure of his own skills and self-sufficiency. So he became unusually skilled in carpentry and blacksmithing and taught these skills to his four sons and some of his slaves.

A remarkably shrewd businessman, Jordan Gatling's estate grew fairly rapidly. He purchased small plots of land and added several small-sized plantations to his original eighty acres. He utilized his woodlands for hog production, grew rice on poorly drained soil, and cleared land for such row crops as peanuts, corn, tobacco and cotton. At the time of his death in 1848 he was one of the larger land owners in Hertford County. His original eighty acres had grown to over twelve hundred acres.

The sons of Jordan Gatling had learned first-hand from their father the value of thrift and hard work. His son, James Henry, at the age of twenty-nine in 1845 had acquired three plantations with a total of 526 acres in Hertford County's Cool Springs District or old District 2, together with eighty acres and numerous lots and buildings near the county seat of Winton. By this time James Henry also had established himself as a well-known wine maker. His wines were marketed widely in eastern Virginia and North Carolina. Jordan's oldest son, Thomas Barnes Gatling, had acquired two plantations in neighboring Northampton County and was the owner of a sawmill. And, before leaving for St. Louis in 1844, Richard Jordan Gatling had acquired a large tract of land lying immediately south of his father's Maney's Neck holdings. The influence of the Gatlings on the region was very pronounced as many planters and farmers attempted to emulate the highly successful farming and business practices of this industrious family.

William, the youngest son of Jordan and Mary Barnes Gatling, soon would travel west, but, following a brief stay in Kansas, he proceeded to Canada. But he did not fare as well economically as did his three brothers. While in Canada he devoted his efforts to developing gold mines at Belleville in Ontario. Despite substantial financial assistance from his brothers, particularly Richard and James Henry, and the formation of a stock company, the gold mining venture did not produce a rich strike. William died of cancer on August 6, 1884 and was buried in the Masonic Lot in Mount Pleasant Cemetery, Toronto, Canada.

His brother, Dr. Richard Jordan Gatling, traveled from his home in Hartford, Connecticut to attend his funeral and administer his estate. Richard's letter of August 9, 1884 to his niece, Rebecca Peebles of Jackson, North Carolina, indicates that William J. Gatling was a man of very modest means. Besides five and one-half acres of land adjoining the mine and a few vacant lots in White Cloud, Kansas, William's assets consisted of a "case of books, gold watch, and a few other artifacts of little value."

In 1824, when Richard Jordan Gatling was six years old, he moved with his parents and brothers and sisters into the two-story frame house that his father had just completed. For many years the Negroes who lived on the plantation referred to Jordan Gatling's new dwelling as the "Great House." This sturdy built farmhouse stood in front of the old log cabin where Richard was born and the new home also fronted upon "a long and level avenue."

Jordan's "Great House" was 36 feet by 30 feet in size with a rear "T" of about twenty feet. The "Great House" had two spacious rooms and a front hallway on the second floor. The first floor consisted of a wainscoted main entrance hallway and a large parlor room on the right. Jordan Gatling's craftsmanship was found throughout the house particularly in the finely crafted mantels that were in the large parlor room and a rear dining room in the "T" addition. The "no-nonsense" design of the chair-railing, window and door trim, and raised paneled doors throughout the house reflected the strength and wisdom of its builder. Shortly after 1900 a two story addition of Victorian design was added to the left end of the front section of the house. By the 1950s this addition was falling in, while Jordan Gatling's original part of the house showed little or no signs of deterioration, save for the lack of proper maintenance. The old log cabin dwelling was used as a kitchen through the Civil War years. When the war ended, some of the former Gatling slaves used it as a dwelling until it burned. The "Great House" was flanked by numerous outbuildings including slave quarters, barns, livestock stables, a blacksmith and carpenter shop, dairy house, and an ice pit for keeping ice cut from the plantation's fish ponds. The ice was kept through the spring well into the summer months. In 1856 Jordan's son, James Henry, constructed a large cotton press on the plantation. Towards the end of the Civil War, James Henry also built a block house to protect himself from a band of renegades who had raided the place in late 1864 and robbed him of about $15,000.00.

Thirty years after Jordan Gatling's death in 1848, his old homestead, under the care of James Henry Gatling, had matured with growing things to a near perfect state. Charles Henry Foster, a Murfreesboro newspaper publisher and frequent visitor to the plantation, printed that the "Great House" and the grounds about it were "a model of scrupulous exact neatness." He observed that "not a leaf or straw or scrap of paper or a particle of any other sort of litter is to be seen anywhere in the whole large yard." The shade trees were numerous and of considerable variety for that section of the country. They included spruces, cedars, cypresses and junipers along with elm, maple and ash. The clumps of boxwood which had been set out many years earlier seemed to have "flourished in vigorous vitality." The "prolific vineyards" which were producing wine as early as 1827 and "thrifty orchards", were producing juice for the making of excellent peach and apple brandy, stood nearby.

Although Jordan Gatling was a substantial slave owner and depended on his Negroes for much of the manual labor on his plantations, he sought to ease the heavy burden and monotony of farm labor. As a matter of fact the first patent holder in the Jordan Gatling family was none other than Jordan himself. He built his own farm implements and patented two of them in 1835. One patent was for a cotton planter and the other patent was for a rotary cultivator. Some thirty years later in 1861 and 1862 Jordan's son,

Richard, would combine the principles of his own cotton planter of dropping off the cotton seed at regular intervals and his father's rotary cultivator to develop the Gatling gun. Instead of dropping seed off at regular intervals, Richard's adaptation of his seed sower and his father's rotary cultivator dropped off bullets at a regular interval. The two patents by Jordan Gatling for farm implements were the first machines patented in the United States for "opening a ridge, sowing and covering cotton seed and chopping (thinning) out cotton plants."

Now past sixty years of age, and with declining health limiting his activity, Jordan Gatling channeled his creative energy into the carving of elaborate walking sticks from dogwood and hickory trees. It is known that Jordan carved at least seven of these fancy walking sticks and at least two are known to survive today. When his son, Richard, reached international prominence, the sticks were placed with an exhibit of Dr. Gatling's relics in a room in the "Great House." Charles Henry Foster stated that the Jordan Gatling walking sticks rivaled "anything possible to the deftest Japanese whittler in delicacy of execution and minute accuracy of design and finish." The walking sticks were covered with native reptiles, fishes, raccoons and other creatures. One stick, about one-half an inch in diameter, was his prized cane. It was covered with poems, essays, and names of history's famous men. It bore a total of 510 words in clearly cut Roman text.

Jordan Gatling during his lifetime had become one of the better known and most widely discussed men in Hertford County, North Carolina. Upon learning of Jordan Gatling's death in 1848, diarist William Darden Valentine of Winton, the county seat of Hertford County, penned that he, "possessed more enterprise than any other man in the county...," then he added, "no one, not one of his neighbors, save his immediate family, regrets his death." The later entry into Mr. Valentine's diary was prompted by the fact that while Jordan Gatling was fiercely independent and extremely resourceful, he was somewhat short on social graces. Basically he kept to himself and bid his neighbors and visitors to do likewise.

Hertford County historian, John Wheeler Moore, wrote years later that, "He [Jordan Gatling] was in many respects a remarkable man. He was full of contrivance, originality and the strangest eccentricities. Charles Dickens would have delighted in his acquaintance and the portrayal of his oddities. While still in vigorous old age he was cut down amid safe predictions on the part of neighbors that they should never look upon his like again."

Charles Henry Foster, the Murfreesboro newspaper publisher, noted that Jordan Gatling was "a man of marvelous thrift — considered somewhat selfish, perhaps in those old, easy going days, but his integrity was never questioned, and he left no heritage of debt to his heirs." But Jordan Gatling's greatest legacy was the fact that he had taught his four sons the merits of honest, hard work. As the years passed, each of Jordan's sons lived up to his father's legacy.

Mother Mary Barnes Gatling was almost thirteen years younger than Jordan Gatling when they married on October 11, 1810. When she died on September 30, 1868, she had survived her husband by some twenty years. The epitaph on her headstone reads, "A devoted wife, an affectionate mother and a true Christian."

Mary Barnes Gatling bore seven children, three daughters and four sons, for Jordan Gatling. Her four sons, including Richard, were educated at Buckhorn Academy, a classical school for boys that stood on the grounds of St. Luke's Episcopal Chapel, now Buckhorn Baptist Church in Como, N.C. Her three daughters Mary Ann, Caroline and Martha had all preceded her husband in death. Caroline lived only twenty-nine days and Mary Ann and Martha reached adulthood. When Jordan Gatling died in 1848, not only had his three daughters died but three of his four sons had left home. Richard was in St. Louis, Missouri, Thomas was in nearby Northampton County and William was in White Cloud, Kansas, later to move to Canada. This left Mary Barnes Gatling at home with a lone son, James Henry, a highly successful planter and well known wine maker. He moved into the "Great House" to manage his mother's large estate along with his own land holdings which were considerable. But James Henry Gatling was about as unsociable as was his late father, Jordan. In later years James Henry's lack of social graces, or at least the proper exercise of them, led to fist fights and hair pulling episodes with some of his neighbors and very nearly cost him membership in Buckhorn Baptist Church. The church incident was caused by the fact that on one Sunday morning James Henry Gatling and his neighbor, Jethro Barnes,

who were squabbling over some runaway hogs, fell to fighting on the church steps. James Henry's unsocial ways also led to his murder in 1879.

In her long widowhood, Mary Barnes Gatling became very active in Maney's Neck affairs. Using her top gig pulled by a beautiful old grey horse, she was often seen visiting friends and the sick, attending Buckhorn Baptist Church, and overseeing her business. Years after her death in 1868, Benjamin B. Winborne, a local historian, described her as "a most estimable woman, very charitable and an angel of mercy and relief to the entire neighborhood." Mr. Winborne, having been reared in Maney's Neck Township, remembered Mrs. Gatling's kindness to his family when he was a young boy.

Only three of the seven children of Jordan and Mary Barnes Gatling were ever married. They were Richard, Thomas B. and Mary Ann. It is reported that in 1841 Richard Jordan Gatling married a woman from Winston-Salem, North Carolina area. But the bride's father strenuously objected and had the marriage annulled. On October 24, 1854 in Indianapolis, Indiana, Richard married Jemima Taylor Sanders of the city where he invented the Gatling gun in 1862. Thomas Barnes Gatling, the oldest son of Jordan and Mary Barnes Gatling, married Rebecca Jane Long about 1830. She was from near Jackson, North Carolina and the couple settled there following their marriage. This marriage produced four children, three girls and a boy. When Thomas Barnes Gatling died on May 14, 1857, he was buried in the Gatling family cemetery on the Jordan Gatling plantation in Maney's Neck Township. His wife, Rebecca Jane Long, was buried beside her husband when she died in 1870. Today there are descendants of Thomas Barnes and Rebecca Jane Gatling residing in northeastern North Carolina, namely in the Jackson, Roanoke Rapids and Elizabeth City areas. On February 9, 1832, Mary Ann Gatling, born September 29, 1813, married John Gatling of the village of Mapleton, located about three miles east of Murfreesboro. John Gatling, born 1807 in Mapleton was the son of David Gatling. John and Mary Ann Gatling had two daughters, Mary and Julia. Many of their descendants today can be found in Bertie and Hertford Counties, particularly in Ahoskie.

In some respects Richard Jordan Gatling seemed like a man without a home particularly during the Civil War. Residing in Indianapolis, Indiana during the war Gatling was falsely accused by some Union leaders as being a Copperhead and secretly selling his gun to the South. Likewise, many Southerners felt betrayed when Gatling sold twelve of his guns to General Benjamin Butler who used them in and around Richmond. The use of the Gatlings by Union forces had little or no impact on the final outcome of the Civil War. In 1866 the federal government adopted the Gatling gun for use in the armed forces and Gatling's fame and fortune grew by leaps and bounds only to come crashing down thirty years later. Ironically, more powerful versions of the Gatling gun today (1993) are standard equipment in United States warships and war planes.

In a March 18, 1873 letter from the Jordan Gatling plantation in Maney's Neck Township to The Norfolk Journal, Charles Henry Foster wrote, "We were shown...a flying machine in embryo, with which when completed an enthusiastic friend expects to navigate the air like any of the feathered tribe." Charles Henry Foster's enthusiastic friend was none other than James Henry Gatling, an older brother of Dr. Richard Jordan Gatling. According to documented eyewitness accounts, James Henry Gatling "was the fool who thought he could fly."

Born July 15, 1816, James Henry Gatling had an obsession with flying objects. As a young boy, he was flying kites of different shapes and designs. He was also looking toward the sky and dreaming of "aeronauts" and flyers. His parents had noticed that in some ways James Henry was different from their other sons. While his brothers were outgoing, he was a quiet, solitary boy possessed with an intense building of his own toys. He made his toys from wood which he carefully shaped, mortised, grooved and assembled with small wooden pegs. The normal toys for boys of his day would have been homemade flutes, sling-shots, rabbit gums, log traps, deadfalls, fish traps and bows and arrows.

Several milestones happened during James Henry's life. When he was eight years old in 1824, he moved into the "Great House" which had replaced the old log cabin as the principal Gatling residence. A few years later a large two story barn was built about three hundred yards west of the "Great House." It was from the loft of this barn that the boy, James Henry Gatling, con-

ducted his early attempts to fly. Although members of his family watched with apparent understanding, the youngster, found himself suddenly the butt of numerous jokes and wisecracks on the Gatling and neighboring plantations by slaves and others.

One of James Henry Gatling's black neighbors, Mrs. Edie Cooper, who was born in 1829 and died in 1923, recalled a number of James Henry Gatling stories in her later years. In one such story she recalled how James Henry Gatling had destroyed an umbrella while trying to parachute from the barn loft. The umbrella "flopped inside out" as the young James Henry Gatling plummeted to the ground. Later, James Henry constructed himself a pair of wings out of fodder (dry corn leaves) and jumped from the same barn loft "flopping" them. He landed hard again, this time badly spraining his ankles.

Not all of James Henry Gatling's early attempts to fly were so ridiculous. Several displayed a remarkable insight and understanding of gravity and aerodynamics. He designed and built a number of wooden frameworks of such models with tiny wooden pegs and covered their wings with paper and made them in the shape of native birds. Approaching manhood, James Henry did not put aside his curiosity for flying things. He was persistent in his efforts to fly because in later years he was convinced that he could build a flying machine to carry human cargo. In the back of his mind he felt certain that such a flying machine would surpass the fame and fortune of his brother Richard's machine gun.

Thus, James Henry Gatling, the dreaming and tinkering boy grew into manhood and became a most successful planter and business man. He came to possess the same self-sufficiency and frugality of his late father, Jordan, and acquired even greater wealth and reputation. Like his father Jordan and famous brother Richard, James Henry Gatling held several patents primarily of agricultural and forest products related.

As the years passed by for James Henry Gatling, he steadily built more than a modest financial empire through farm enterprises, wine-making, store-keeping and land speculation. While his personal fortune was growing so was his reputation as an eccentric and odd-ball, fiery-type character. Pursued by a number of women, James Henry remained a bachelor because he believed, "that marriage was a good thing but a wise man ought to consider of it all of his life." In 1868 he did make one marriage proposal, but the lady of his choice never gave him an answer. While marriage may not have been a serious matter with James Henry, flying was throughout his adult life as he was constantly tinkering with flying and vowed to put "a flying machine in the air before I die."

The older James Henry Gatling grew, the hotter his temper grew. This was particularly true with reference to people who poked fun at his attempts to fly. One particular incident happened when James Henry was 57 years old and a neighbor, James Majette, laughed at his attempts to fly. It seems the two men got into a fight which turned into a royal hair pulling affair with James Henry getting the worst of the hair pulling. Yet another fight occurred between James Henry and another neighbor on the steps, of all places, of the Hertford County Courthouse in Winton. James Henry also appeared to have gotten the worst end of this fight as he had "first struck Mr. Brett a light blow with a heavy stick whereupon Mr. Brett wrestled the stick from Gatling and gave him several severe cuts with it about the head, making the blood flow profusely. They were immediately taken into Court and fined twenty dollars each for contempt of court. They will be presented by the Grand Jury at the next term for a breach of peace." Later, Charles Henry Foster, the Murfreesboro newspaper editor, published that the two "combatants were placed on good behavior."

In the spring of 1873 James Henry Gatling had indeed constructed a flying machine and was waiting for just the right day to fly the thing. One Sunday in May 1873 when James Henry returned home from services at nearby Buckhorn Baptist Church, he noticed that a brisk breeze was blowing and this was the day he had been waiting for to fly "The Old Turkey Buzzard" which was the name that the locals had stuck on his flying machine. Some days earlier the flying machine had been hoisted high on top of the cotton press with the intent of having it pushed off and on its way. James Henry offered to let his second cousin, Roschus Gatling, make the first flight, but Roschus declined, "you built the damn thing, now you fly it!" Across the plantation James Henry had laid out a one-mile course which he hoped to fly over when he took off and became airborne.

In addition to Roschus Gatling, others present for the flight of "The Old Turkey Buzzard" were Euclid Howell, George Banks, Asbury Cooper, Issac Jordan, and Miss Matilda Bassett, James Henry Gatling's housekeeper. Liza, the mulatto helper of Miss Bassett also was present. With its nose in the breeze, with its propellers roaring full blast and with James Henry Gatling at the controls, he motioned for the Negroes to push the flying machine off the top of the cotton press. The flying machine took off and went about a hundred feet before crashing into a hefty-sized elm tree and falling to the ground. Years later Stella Gatling Howell quoted her father, Roschus Gatling, as saying that James Henry was only slightly injured in the crash, but the flying machine had suffered a wrecked wing and other damages, some serious.

News of James Henry Gatling's attempts to fly spread like wildfire throughout eastern North Carolina and eastern Virginia and was talked about for many years among the locals. While many people admired James Henry Gatling for attempting to fly, others continued to ridicule him, thus making his touchy disposition even worse. Somewhat discouraged, but still hopeful that he would be the first to fly, James Henry placed the flying machine in storage under the shed of the cotton press.

Six years later with the flying machine still in storage and with his dream to fly intact, James Henry Gatling was brutally murdered on September 2, 1879. The murderer was William H. Vann, a Murfreesboro Township man who, on the day before, had been refused a carriage ride by James Henry. James Henry's refusal to give Vann a ride had enraged Vann so that he had threatened to kill James Henry. James Henry apparently did not take the threat seriously, but early the next morning Vann slipped quietly on to the Gatling plantation and found James Henry down at the hog pen feeding his hogs. There Vann shot him in the head and clubbed him. Roschus Gatling, James Henry's second cousin and a Hertford County constable, prevented outraged citizens from killing Vann right on the spot.

In October, 1879, William Vann's trial took place in Winton at the Hertford County Courthouse. This highly celebrated trial was so large that most of the businesses in the county closed for the duration of the trial. Among those present when the death sentence was pronounced on William Vann, was Dr. Richard Jordan Gatling. However, Vann's death sentence was commuted by Governor Thomas J. Jarvis and a few years later Vann was out of prison living with his brother.

At an auction held at the Jordan Gatling plantation some months following James Henry's death, no one expressed an interest in buying his old flying machine. The flying machine remained stored under the shed of the cotton press until 1905 when the cotton press and it were destroyed by fire. The fire was witnessed by this writer's grandmother who, two days before, had seen the flying machine in storage in the cotton press.

The murder of James Henry Gatling not only snuffed out the life of a most remarkable man, but ended his lifelong dream of flying. Twenty-four years later, and one hundred miles east at Kitty Hawk, North Carolina, the Wright Brothers would accomplish what James Henry Gatling had attempted on a farm in Hertford County, North Carolina.

Although James Henry Gatling may have been an unsociable character at times, his death hit the Gatling family members extremely hard. William Jesse Gatling, the younger brother expressed the family's deep sorrow over James Henry's death in a September 31, 1879 letter from his home in Canada to his niece, Rebecca Gatling Peebles in Jackson, North Carolina, "I hope your mind is getting soothed by time of your troubles — time alone can bind up the broken heart — your troubles are truly great. Brother Henry was very, very dear to me. Never did one have so kind a brother — and all or ever my desire for success in some sort, his great kindness — but alas, we shall never see him again — nor the kind face of Mr. Peebles, your late husband. Life is a sad dream, but a reality and we must accept its events as best we can and with the greater the resignation possibly the better."

With the death of James Henry Gatling, the burden of settling his estate and disposing of the North Carolina homeplace fell primarily on the shoulders of Dr. Richard Jordan Gatling. Dr. Gatling, who was residing in Hartford, Connecticut, at that time, was at the height of his gun manufacturing career. His only other surviving brother or sister was brother William Jesse Gatling who was living in Canada, but did not have the resources to travel and stay in North Carolina for any length of time.

Suffering from a life threatening case of smallpox and left for dead for three months in a pest house in Pittsburgh in the winter of 1845-46, Richard Gatling regained his strength and decided to study medicine so he could take care of himself and not practice. In 1847-49 he attended medical college in Indiana and Ohio and graduated in medicine in 1849 from Ohio Medical College in Cincinnati. From that time on he was known as Doctor Gatling and wore a beard to cover the pock-marks on his face.

It is not known how many times Dr. Richard Jordan Gatling returned to his native Hertford County after he left in 1844. There are no documented accounts that Richard's wife, Jemima, ever accompanied her husband on any of his trips back to his native county in North Carolina. She did travel with him on several of his trips to Europe and Russia in the 1870s and 1880s. A number of Richard's visits, particularly after he became famous, were recorded by local diarists and local newspapers. Nearly all of his earlier visits had been made under happier circumstances and he was always well received by the local folk who gathered around him at Buckhorn Baptist Church like he was a celebrity. But Richard Jordan Gatling would find the task of settling his brother's estate and that of selling the homeplace to be extremely difficult and most painful. During the process, he made a number of trips to Hertford County staying mainly at the old homeplace. But on at least two trips Dr. Gatling stayed in Jackson with his niece, Rebecca Gatling Peebles and in Murfreesboro at the elaborate home of Congressman Jesse J. Yeates. Dr. Gatling had taught the Congressman when he was a teacher in the late 1830's in an old field school in Hertford County. Richard Jordan Gatling was extremely fond of his niece, Rebecca, who was the daughter of Thomas Barnes Gatling. She and her famous uncle exchanged a bevy of letters through the years. These letters, many of which survive today, substantiate the fact that the gun's inventor was a compassionate and mild-mannered man who had a deep attachment and strong feeling for his North Carolina relatives and roots.

Due to heavy business commitments and trips to Europe, the process of selling the homeplace and settling his brother's estate dragged on for almost ten years. Dr. Gatling relied primarily on

Murfreesboro attorney, B. B. Winborne to handle the day to day details of closing the estate. Since William Jesse Gatling was heavily indebted to his late brother, James Henry, for a sum of "several thousands of dollars" he relinquished any claims on James Henry's estate and on the homeplace. This left the only direct Gatling heirs of James Henry Gatling to be his brother, Dr. Richard Jordan Gatling and a niece, Rebecca Gatling Peebles and a nephew, Isaac Gatling, the surviving children of brother Thomas Barnes Gatling. According to local records, the estate of James Henry Gatling was worth a sum in excess of $50,000 which was a large amount in those days.

Several small parcels of land and timber tracts along with a large sawmill were sold off in late 1879 and 1880. In early 1881 a deal was struck with E. G. Sears to purchase the homeplace for $2200 with the purchase price to be spread out over three years. Shortly thereafter Dr. Gatling held an auction sale at the homeplace to dispose of any household furnishings and farm implements not claimed by immediate family members and close cousins. Before the auction sale, the Gatling family portraits were given to Rebecca Gatling Peebles, of nearby Northampton County, where they remain today with her heirs. The six portraits were painted in the early 1840s by Oliver Perry Copeland who traveled throughout the region painting the prominent families. Other items such as Jordan Gatling's beautiful walking sticks and family business papers were given to close cousins including Roschus Gatling who was residing at Grandfather James Gatling's old homeplace. It was during this time that Dr. Richard Jordan Gatling gave his North Carolina Bible to his second cousin, Roschus B. Gatling. Dr. Gatling had purchased the Bible in the early 1840s when he ran a small store at Frazier's Crossroads in Hertford County and had recorded his date of birth in it. On December 8, 1881 he gave the Bible to Roschus and inscribed the front inside cover with that date and the following, "Presented to R. B. Gatling by R. J. Gatling to remain in the family as a family keepsake forever." Today the Bible is in the possession of Roschus' great-grandson, Harold Gatling of Murfreesboro, North Carolina. Before the auction, Dr. Gatling claimed "the two large mahogany tables in the parlor of the old home as mementoes of the old home." Later he had the two tables shipped by steamer to his home in Hartford, Connecticut.

The sale of the old Jordan Gatling homeplace in 1881 did not mean that James Henry Gatling's estate had been completely settled. His estate was so extensive that efforts to sell all of the individual parcels contained in it, were not successful until 1888 when the last parcel of land, a lot that James Henry owned in Des Moines, Iowa was finally sold. One of the more interesting aspects of James Henry Gatling's extensive estate were some lots contained in a town he had surveyed and laid off in 1873 just outside of the county seat of Winton. His new town bore the name of Gatlingburg and had a main street ninety feet wide and a mile long. There were four streets that ran parallel with main street and eleven cross streets, all named. The town was divided into 480 building lots 150 feet by 75 feet. Like flying, James Henry Gatling's new town of Gatlingburg was destined to become one of his dreams which was never realized. Records in the Hertford County Register of Deeds Office indicates that only a few of the lots were ever sold and eventually the town site was absorbed by its neighbor, Winton. It has been reported down through the years that one of Jordan Gatling's sons was responsible for the founding of Gatlingburg, Tennessee. While there is no evidence to support this theory, it is conceivable because Richard, on his way west to St. Louis and William, on his way west to St. Cloud, Kansas, went through the Gatlingburg, Tennessee region.

The estate of Richard's brother James Henry had not been completely settled when Richard received "an urgent dispatch from Dr. Clark of Toronto, Canada informing me that Bro. William was in a dying condition." By the time that Dr. Gatling arrived in Toronto on August 8, 1884, he learned that his last surviving brother had died two days earlier. He was deeply saddened by his brother's death as he wrote in an August 9, 1884 letter to his niece, Rebecca Gatling Peebles of Jackson, North Carolina, "It is needless to say his death has filled my heart with sorrow and grief; but, we must all sooner or later pass from earth. I attended to the sad duty of burying dear brother. I had a nice coffin made for brother and I bought some nice wreaths of flowers which were placed on the coffin and on the grave. Some friends also sent flowers. I had a Minister of the Church of England read the Episcopal burial service at the grave which was very beautiful and impressive."

Thus, the great inventor had outlived all of his brothers and sisters, but this did not end his contact with his native county and area of North Carolina. He continued to correspond with his niece, Rebecca Gatling Peebles and remained in close contact with his second cousin, Roschus B. Gatling, who continued to live at Grandfather James Gatling's old homeplace.

The sadness in Richard Jordan Gatling's family would continue because in 1890 his beloved niece, Rebecca Gatling Peebles, died. Three years later, in 1893, Richard wrote his second cousin, Roschus, that his wife's health was not good and that his business was in "a serious state of decline."

Personal family tragedy was nothing new to Dr. Gatling because he and his wife, Jemima Taylor Sanders, had lost two children early in their marriage. Their first daughter, Mary, was born in 1855, died when she was five years old. In 1861 their first son, William Sanders Gatling, was born but died two years later.

Hobbled by mounting health problems and financial woes, Dr. and Mrs. Gatling moved in 1897 from Hartford, Connecticut to New York City to live with their daughter, Ida Gatling Pentacost. Their two sons, Richard Henry and Robert B. Gatling, were already living in New York City. Robert was engaged in fire insurance business while Richard Henry was working in real estate.

Richard and Jemima Gatling clearly did not wish to leave their large Italianate mansion at 27 Charter Oak Place in Hartford, Connecticut in 1897. Hartford had been their home for twenty-seven years and they had many friends living there. Likewise, their friends in Hartford deeply regretted to see the Gatlings leave their city. Rev. Frank Dixon, Pastor of Hartford's South Baptist Church, where the Gatlings had faithfully attended, perhaps best expressed Hartford's sorrow over the departure of the Gatlings in a November 22, 1897 letter to Mrs. Richard Jordan Gatling at her daughter's home in New York City, "If you know how badly we miss you from the church and from our lives your heart would incline you to come back to Hartford I know."

So far as can be ascertained, Richard and Jemima Gatling never returned to Hartford, Connecticut. Sadly, the Gatlings sold their beautiful home overlooking the massive Colt Company's firearms manufacturing plant where the Gatling gun was manufactured. The Gatlings had purchased the house in 1870 from Mr. Kingsbury, a cashmere manufacturer who had the house built two years earlier. Following a succession of owners, by 1976 the house had been condemned by the City of Hartford as unfit to live in. Today, the old home of Richard and Jemima Gatling has been magnificently restored by the Gatling Mansion Limited Partnership of Hartford.

Living in New York City was not easy for Dr. Gatling and finally the urge to go west struck the old inventor once again. In late 1901 Dr. and Mrs. Gatling moved from New York City to St. Louis, Missouri, the place where Dr. Gatling had first settled in 1844 after leaving home in North Carolina. Ironically, this also was the place where Dr. Gatling had made his first substantial money by selling seed sowers and wheat drills. This time around Dr. Gatling was hoping to strike it rich again through agricultural related implements and equipment. He had recently founded the Gatling Motor Plow Company and was busily perfecting a few business formalities prior to marketing his new motor plow which in reality was a tractor. This time however, fate would intervene as for the past three years Dr. Gatling had been a victim of serious heart trouble. His heart condition was suddenly complicated by a severe case of the grip. So with little or no money left, and with both Richard's and Jemima's health in serious question, their son, Richard Henry Gatling of New York City, went to St. Louis in early February, 1903 and took them back to New York to live once again with their daughter, Ida.

Within just a few weeks, Dr. Richard Jordan Gatling died on February 26, 1903. The world, and particularly his native county and township, mourned the loss of this great but sensitive man. Following services in New York City, his body was taken to Indianapolis, Indiana by train. Richard Jordan Gatling's remains arrived in Indianapolis over the Big Four, at 11:45 a.m. on March 2nd and were taken to the home of John R. Wilson at 1308 Central Avenue. Mrs. Richard Jordan Gatling, her daughter Ida Gatling Pentecost and her son Richard Henry Gatling had accompanied the remains from New York City. Funeral services conducted by Tutewiler Funeral Home, were held at 2 p.m. at the Wilson residence. The funeral address was delivered by the Rev. A. R. Benton, who had been an acquaintance of Dr. Gatling's for nearly fifty years. The pall bearers were John S. Duncan, John R. Wilson, David Wallace, Walter Golt, Robert P. Duncan and Judge James Leathers. Burial was in plot number 9 of lots 9 and 10 of Section C in Crown Hill Cemetery. The lots had been purchased on March 1, 1865. Mrs. Richard Jordan Gatling and her son and daughter returned to New York City two days after the funeral.

The bodies of the two deceased young children of Richard and Jemima Gatling, William and Mary, were moved to this plot from another location. Later, the body of their son Robert B., who was born on October 9, 1872, and died, December 2, 1903, was buried here. The body of the Gatling's long time cook and housekeeper, Rachel Stepney, a black woman who died in 1885 was buried here over the strenuous objections of the cemetery's management. John Waters Gatling, Dr. Gatling's grandson, states that his grandfather's prominence prompted local officials to accept his wishes that the body of his cook and housekeeper be buried in his family's plot.

When Mrs. Richard Jordan Gatling died on September 26, 1908 in New York City, her body was laid to rest in Crown Hill Cemetery. Because Dr. Gatling was virtually penniless when he died in 1903, his wife, Jemima, spent her last years living in New York with their daughter, Ida, and being supported by modest financial gifts from her son, Richard Henry Gatling. Around 1915, Richard Henry Gatling had the large Gatling monument in Crown Hill cemetery erected. Ida Gatling Pentecost, daughter of Richard and Jemima Gatling died on January 13, 1911 and was buried in Brooklyn, New York. Her brother, Richard Henry Gatling died on January 14, 1941 in St. Petersburg, Florida and was cremated by his second wife.

Only two of Richard and Jemima Gatling's children, Ida and Richard Henry, had any children. Ida had one daughter, Ida Marguerite Pentecost, who married Albert C. Newcombe. There are a number of descendants of Ida and Albert C. Newcombe living today. Richard Henry Gatling had three children, Addison, Rosalind and John Waters Gatling. John Waters Gatling, a grandson, who resides in Asheville, North Carolina is the most direct descendant of Dr. Richard Jordan Gatling living today. John, along with his late brother, Addison will leave no direct descendants since neither had any children. There are however, several grandchildren of Rosalind Gatling Hawn living today, primarily in the the northeast.

When Richard Jordan Gatling's second cousin, Roschus B. Gatling died in 1909, he left five daughters and two sons. The descendants of these five daughters and two sons are numerous and bear the names of Gatling, Taylor, Kitchen, Howell, Bowles, Hill and Parker. While most of these Gatling cousins are widely scattered throughout eastern North Carolina and eastern Virginia, a good number of them reside in the home area of Hertford and Northampton Counties. It is through these Gatling cousins that much of the artifacts and archival materials relating to Jordan, Richard Jordan, and James Henry Gatling survive today. They are generous and kind people.

It deeply saddens this writer when he rides by the old Jordan Gatling plantation in Maney's Neck Township in Hertford County, North Carolina. Approaching the old plantation via "Gatling's Avenue", a mile long road which was originally built as a private racetrack for the Gatling horses and today is a paved public road connecting US 258 and NC 1310, there remains little of the old plantation except the family cemetery. In the past 25 years the old plantation has gone through a succession of owners, who bulldozed down the old Jordan Gatling house in 1979 and did extensive clearing and bulldozing work. In the process, the old fish ponds and building sites were destroyed. The old plantation which has not been farmed for several years is gradually being sold off for mobile home lots.

Between 1963 and 1966, I attempted without success to purchase the old Jordan Gatling house. It had been a longtime dream of mine to restore the house for my home. But the owners back then simply refused any offer from anyone to buy the house and a small amount of land surrounding it along with a access route. Although my own effort to purchase, save, and restore the old Gatling house had failed, the owners did agree after some time passed to sell the interior woodwork from the large parlor room on the first floor of the Gatling house to The Murfreesboro Historical Association, Murfreesboro, North Carolina to use as the focal point of a Richard Jordan Gatling museum in Murfreesboro. The Gatling room was reassembled in the restored

1790 William Rea Store which was being adapted for use as a Murfreesboro museum featuring exhibits on Indians, agriculture, education, river transportation and Dr. Richard Jordan Gatling. Today, the restored Gatling Room houses exhibits on Dr. Gatling, Gatling family memorabilia, and an authentic Gatling gun.

During World War II, the federal government honored Richard Jordan Gatling by naming two ships, a destroyer and a Liberty ship after him. Unfortunately, his native state, North Carolina, has not been so generous with its recognition. An obscure and incorrect highway marker located a half mile south of the old Jordan Gatling plantation on US 258 in Maney's Neck Township represents North Carolina's only official memorial to Richard Jordan Gatling and his inventive family. How tragic because this creative and industrious family held over fifty patents, seriously flirted with a screw propeller and a flying machine, and gave the world the ultimate weapon of the 19th century, the Gatling gun. Although North Carolina's official recognition of Dr. Gatling is scant, the North Carolina relatives of Richard Jordan Gatling today fondly and proudly remember him as "Cousin Dick" Gatling.

Cover Photographs

Front Cover – Front end view of a 1903 30 caliber Gatling gun (Number 1155) owned by The Historial Association, Murfreesboro, North Carolina. The gun was obtained for The Murfreesboro Historical Association from Edward J. Schultheis of Glendale, New York by E. Frank Stephenson, Jr., in October 1967. Photograph by E. Frank Stephenson, Jr., June 1985.

Front Inside – Tin type taken of Richard Jordan Gatling in Indianapolis, Indiana by Mart. L. Ohr in 1854. Photograph courtesy of John Waters Gatling, Asheville, North Carolina.

Back Inside – Skinner's Gut on the Meherrin River where Richard Jordan Gatling perfected a screw propeller some months before Erricson was granted a patent for same. Skinner's Gut is located about three-fourths of a mile from the Jordan Gatling plantation where Richard Jordan Gatling grew up. When Richard Jordan Gatling had perfected his screw propeller at Skinner's Gut on the Meherrin River and on the fish ponds on his father's plantation, he asked his father for permission to go to Washington to obtain a patent for his invention. But his father thought it was foolish for him to go to Washington so he refused to let his son go. Finally Jordan Gatling did grant his son permission to go and when Richard arrived at the patent office in Washington he found that Erricson had been issued a patent for the screw propeller three days earlier. Photograph by Colbert P. Howell and E. Frank Stephenson, Jr., 1970.

Back Outside – Floral wreath in relief encircling the name GATLING on the obelisk at the Jordan Gatling Cemetery in Maney's Neck Township (Como), Hertford County, North Carolina. Photograph by Colbert P. Howell and E. Frank Stephenson, Jr., 1970.

Eng^d by Geo E Perine, N.Y.

Yours truly R J Gatling

Richard Jordan Gatling. Steel engraving appearing in the front inside cover of <u>Science Record</u>, 1872.

Early model Gatling gun. Photograph courtesy of C. B. Robertson Family, Jackson, North Carolina.

Early model Gatling gun. Photograph courtesy of C. B. Robertson Family, Jackson, North Carolina.

17

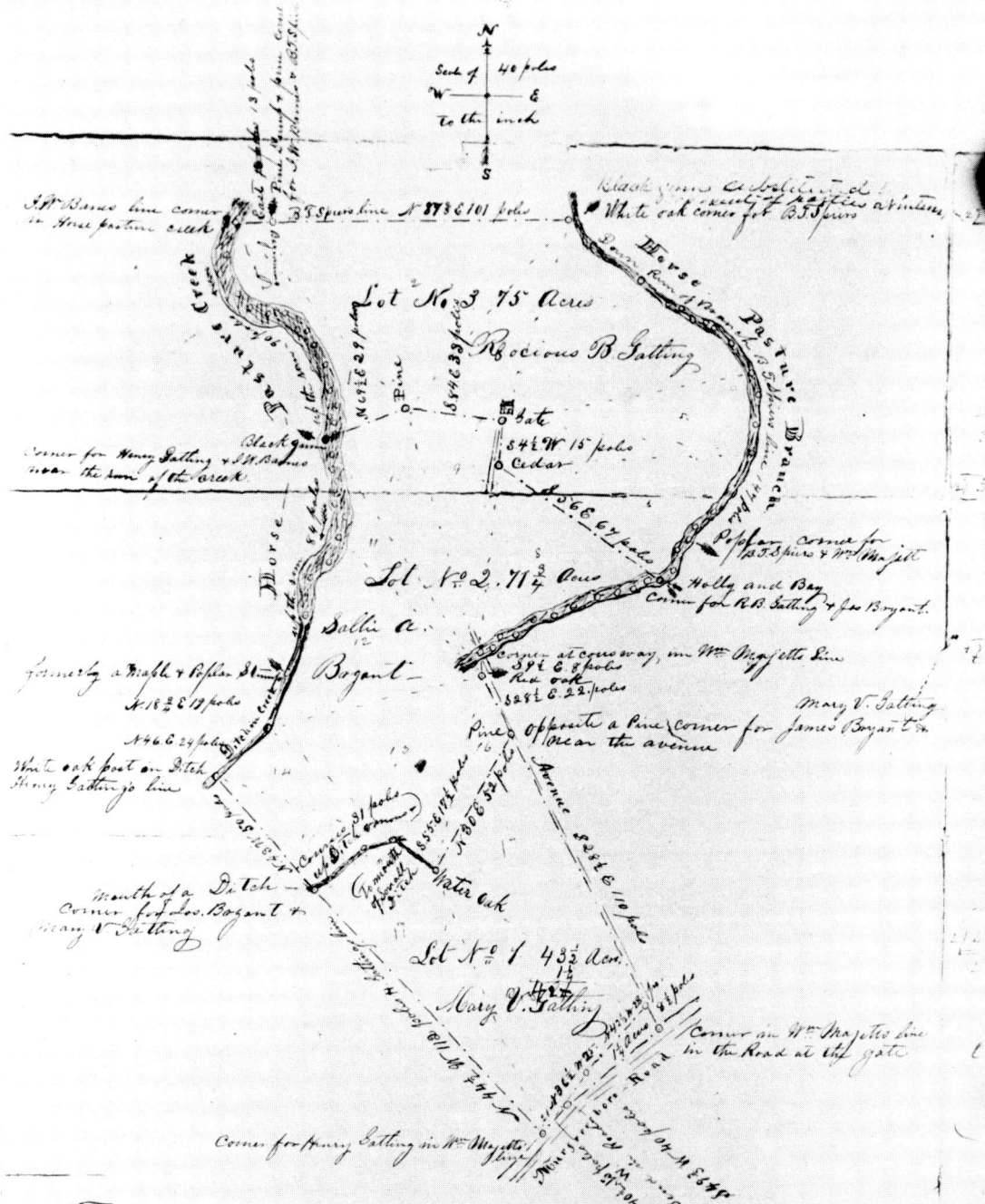

Platt of the lands of Richard Jordan Gatling's grandfather, James Gatling in Como, North Carolina. Richard Gatling's father, Jordan Gatling was born on this plantation. Courtesy of Mrs. Elsie Taylor Parker, Murfreesboro, North Carolina.

18

On September 12, 1818, Richard Jordan Gatling was born in this log cabin. He spent the first six years of his life in the cabin before his family moved into the two-story "Great House" in 1824. From Potter's American Monthly, May 1879.

Chimney base to the log cabin where Richard Jordan Gatling was born. Photograph by E. Frank Stephenson, Jr., 1960.

Jordan Gatling
Son of James and Mary Cowper Gatling
Father of Richard Jordan Gatling
b. May 14, 1783   d. April 12, 1848
Burial: Jordan Gatling Cemetery, Como,
North Carolina. Portrait was painted circa
1842 by Oliver Perry Copeland. Courtesy
C. B. Robertson Family, Jackson, North
Carolina.

Mary Barnes Gatling
Mother of Richard Jordan Gatling
b. October 30,1795   d. September 30,
1868
Burial: Jordan Gatling Cemetery, Como,
North Carolina. Portrait was painted circa
1842 by Oliver Perry Copeland. Courtesy
of C. B. Robertson Family, Jackson, North
Carolina.

Thomas Barnes Gatling
Brother of Richard Jordan Gatling
b. September 29, 1811   d. May 14, 1857
Burial: Jordan Gatling Cemetery, Como,
North Carolina. Portrait was painted circa
1842 by Oliver Perry Copeland. Courtesy
C. B. Robertson Family, Jackson, North
Carolina.

James Henry Gatling
Brother of Richard Jordan Gatling
b. July 15, 1816   d. September 2, 1879
Burial: Jordan Gatling Cemetery, Como,
North Carolina. Portrait was painted circa
1842 by Oliver Perry Copeland. Courtesy
C. B. Robertson Family, Jackson, North
Carolina.

Martha Sarah Gatling
Sister of Richard Jordan Gatling
b. September 29, 1828 d. July 22, 1846
Burial: Jordan Gatling Cemetery, Como, North Carolina. Portrait was painted circa 1842 by Oliver Perry Copeland. Courtesy C. B. Robertson Family, Jackson, North Carolina.

William Jesse Gatling
Brother of Richard Jordan Gatling
b. July 19, 1826 d. August 6, 1884
Burial: Mount Pleasant Cemetery, Toronto, Canada. Photograph courtesy of Mr. Robert Lee Gatling, Ahoskie, North Carolina.

Mary Anne Gatling
Sister of Richard Jordan Gatling
Photograph courtesy of Mr. Robert Lee Gatling, Ahoskie, North Carolina.

Fan belonging to Martha Gatling, sister of Richard Jordan Gatling. Courtesy Miss Rebecca Gatling Long, Jackson, North Carolina.

John Gatling, husband of Richard Jordan Gatling's sister, Mary Anne Gatling. John Gatling was from Mapleton in Hertford County, North Carolina and a cousin of the Maney's Neck, North Carolina Gatlings. Photograph taken in Norfolk, Virginia by Walter's Photography Gallery on 47 East Main Street. Courtesy Mr. Robert Lee Gatling, Ahoskie, North Carolina.

Richard Jordan Gatling
Inventor of Gatling Gun
b. September 12, 1818   d. February 26, 1903
Burial: Crown Hill Cemetery, Indianapolis, Indiana. Portrait was painted circa 1842 by Oliver Perry Copeland (Restored 1990
by Colbert P. Howell, Raleigh, North Carolina). Courtesy C. B. Robertson Family, Jackson, North Carolina.

Artist's sketch of James Henry Gatling and his flying machine. Sketch by Bill Ballard, Raleigh, N.C., Courtesy of F. Roy Johnson, Murfreesboro, North Carolina.

James Henry Gatling, Brother of Richard Jordan Gatling. Photograph taken in Petersburg, Virginia by G. W. Mannis' Photographic Gallery. Date unknown. Photograph courtesy of Mr. Robert Lee Gatling, Ahoskie, North Carolina.

William Jesse Gatling, brother of Richard Jordan Gatling. b. July 19, 1826 d. August 6, 1884. Burial: Mount Pleasant Cemetery, Toronto, Canada. Photograph courtesy Mr. Robert Lee Gatling, Ahoskie, North Carolina.

23

Thomas Barnes Gatling
Brother of Richard Jordan Gatling
Tin type taken in Petersburg, Virginia. Courtesy Mr. John Waters Gatling, Asheville, North Carolina.

Brick dated 1824 in the chimney of the Jordan Gatling house in Como, North Carolina. Photograph by E. Frank Stephenson, Jr., February 15, 1965.

1870 photograph of Jordan Gatling plantation house, Como, North Carolina. The house was built by Jordan Gatling in 1824. Photograph courtesy of Miss Rebecca Gatling Long, Jackson, North Carolina.

1915 view of Jordan Gatling plantation house, Como, North Carolina. Photograph by R. H. Worthington, Windsor, North Carolina.

1920 view of Jordan Gatling plantation house, Como, North Carolina. Photograph by Richard McGlohon, Winton, North Carolina.

1920 view of Jordan Gatling platation house, Como, North Carolina. Photograph by Richard McGlohon, Winton, North Carolina.

1956 view of Jordan Gatling plantation house, Como, North Carolina. Photograph by E. Frank Stephenson, Jr.

1966 view of Jordan Gatling plantation house, Como, North Carolina. Photograph by E. Frank Stephenson, Jr.

Jordan Gatling plantation house, Como, North Carolina. Photograph by E. Frank Stephenson, Jr., on December 29, 1965.

Jordan Gatling plantation house, Como, North Carolina. Photograph by E. Frank Stephenson, Jr., on April 21, 1968.

1969 view of Jordan Gatling plantation house, Como, North Carolina. Courtesy North Carolina Department of Cultural Resources, Raleigh, North Carolina.

1969 view of Jordan Gatling plantation house, Como, North Carolina. Courtesy North Carolina Department of Cultural Resources, Raleigh, North Carolina.

1972 view of Jordan Gatling plantation house, Como, North Carolina. Photograph by Colbert P. Howell and E. Frank Stephenson, Jr.

Architectural detail on Jordan Gatling plantation house, Como, North Carolina. 1972 photograph by Colbert P. Howell and E. Frank Stephenson, Jr.

Architectural detail on Jordan Gatling planta-tion house, Como, North Carolina. 1972 photo-graph by Colbert P. Howell and E. Frank Stephenson, Jr.

31

End view of Jordan Gatling plantation house, Como, North Carolina. 1972 photograph by Colbert P. Howell and E. Frank Stephenson, Jr.

NORTH CAROLINA.

Comptroller's Office, January, 1840

To all to whom these Presents may come, Greeting:

Richard Gatling

hereby authorised to sell GOODS, WARES AND MERCHANDIZE, for one year, and no longer, commencing from the first day of April, 1840, having paid six ———— Dollars, agreeably to an Act of the General Assembly of 1822, chapter the first.

Thos. Colley *Comptroller.*

Countersigned by Richard J. Cowper Sheriff of Hertford county, this 2nd day of September 1840

R.J. Cowper Sheriff
By M.L. Jernigan DS

Business license granted for Richard Jordan Gatling in January 1840 to operate a store in Hertford County. He operated a store at Frazier's Crossroads near Union and later at Winton. Gatling also taught school in an old field school in Hertford County in the early 1840s. Courtesy Mrs. Elsie Taylor Parker, Murfreesboro, North Carolina.

33

NO PRINTED SPECIFICATIONS AVAILABLE

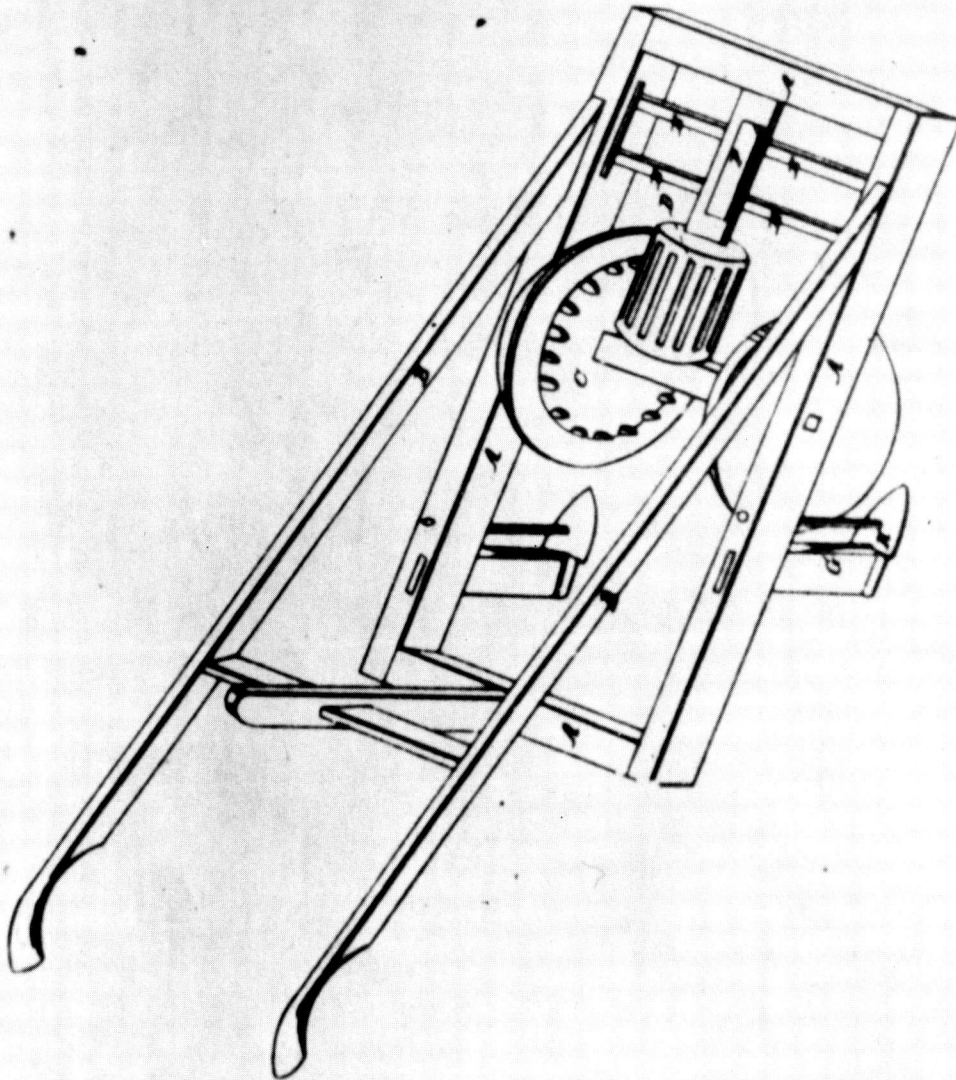

Drawing for Jordan Gatling's rotary cultivator, patented June 19, 1835. Courtesy National Archives, Washington, D. C.

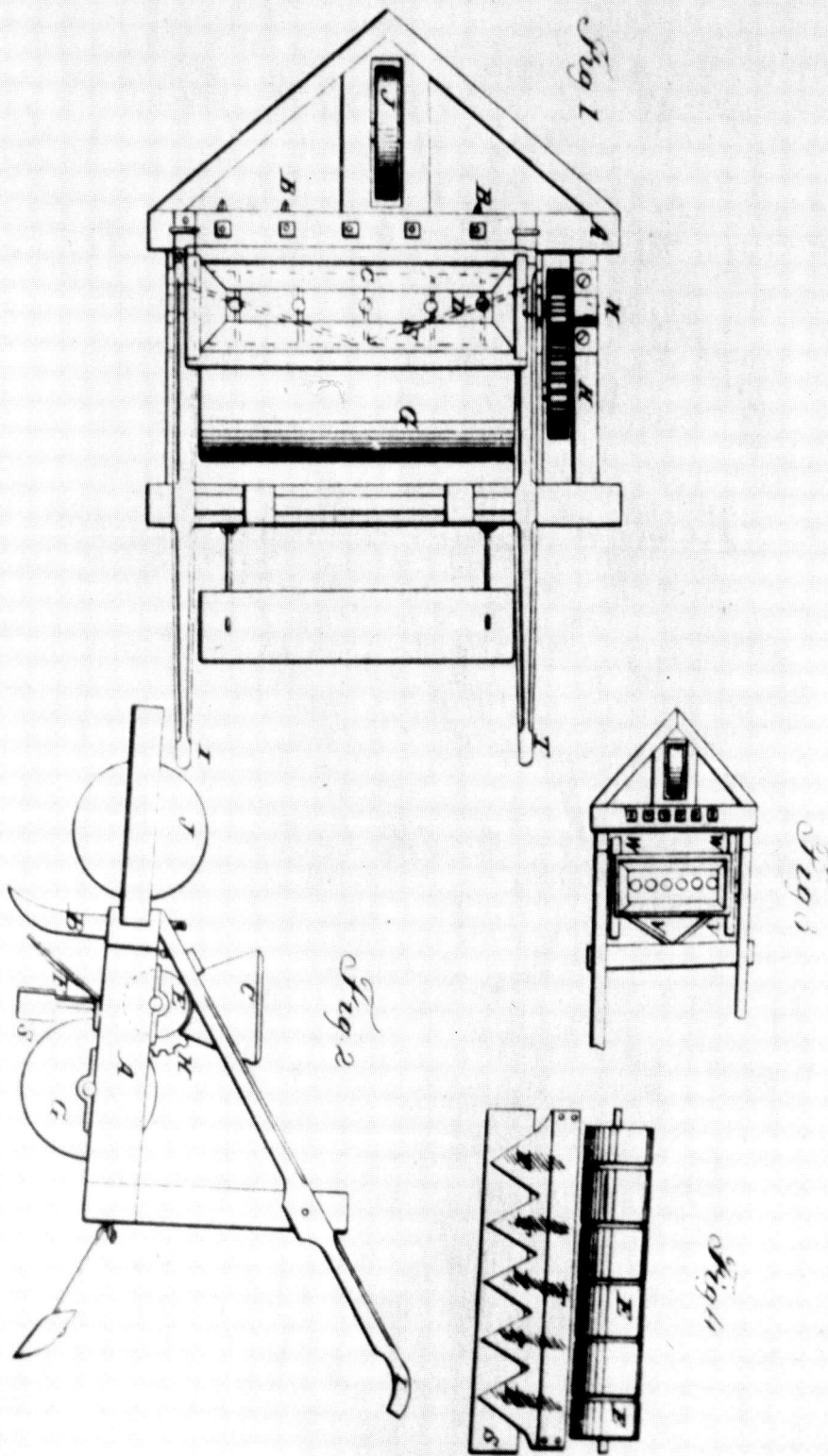

Drawings for Richard Jordan Gatling's seed sower, patented May 10, 1844. Shortly after this patent was granted, Gatling headed west where he settled in St. Louis, Missouri. Here he converted this seed sower into a wheat drill that made him a wealthy man. Drawings courtesy National Archives, Washington, D.C.

Tin type taken of Richard Jordan Gatling in Indianapolis, Indiana by Mart. L. Ohr in 1854. Courtesy Mr. John Waters Gatling, Asheville, North Carolina.

Tin type taken of Jemima Taylor Sanders in Indianapolis, Indiana on the eve of her marriage in 1854 to Richard Jordan Gatling. Tin type by Mart. L. Ohr. Courtesy Mr. John Waters Gatling, Asheville, North Carolina.

Richard Henry Gatling, son of Richard and Jemima Gatling. Courtesy Mr. John Waters Gatling, Asheville, North Carolina.

Robert Boone Gatling, (October 9, 1872 – December 2, 1903) son of Richard and Jemima Gatling. Courtesy Mr. John Waters Gatling, Asheville, North Carolina.

Ida Gatling, daughter of Richard and Jemima Gatling. Photograph taken when Ida was 14 years and 5 months old by Horace L. Bundy, Photographic Portraits, No. 370, Main Street, Hartford, Connecticut. Courtesy Mr. Robert Lee Gatling, Ahoskie, North Carolina.

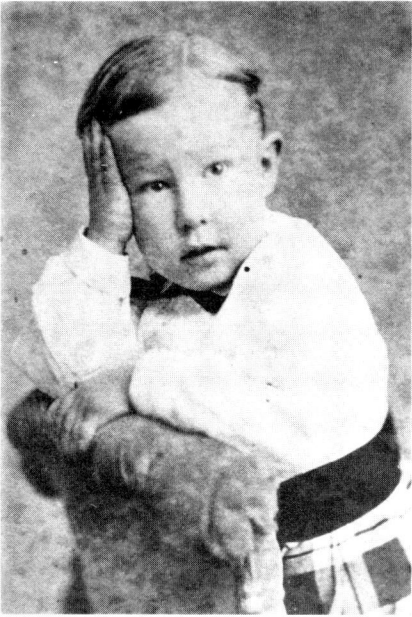

Robert Boone Gatling, son of Richard and Jemima Gatling. He was two years and eleven months old when the photograph was taken by Bundy Photographer, Hartford, Connecticut. Courtesy Mr. Robert Lee Gatling, Ahoskie, North Carolina.

Richard Henry Gatling (March 7, 1870 – January 14, 1941), son of Richard and Jemima Gatling. Courtesy Mr. John Waters Gatling, Asheville, North Carolina.

Rachel Stepney, Richard and Jemima Gatling's longtime nurse. She was born 1815 and died in Hartford, Connecticut on March 30, 1885. Burial was in Crown Hill Cemetery, Indianapolis, Indiana. Courtesy John Waters Gatling, Asheville,

Robert Boone Gatling, son of Richard and Jemima Gatling. He was two and one-half years old when the photograph was taken by R. S. DeLamater Photographer, Hartford, Connecticut. Courtesy Mr. Robert Lee Gatling, Ahoskie, North Carolina.

Richard Henry Gatling, son of Richard and Jemima Gatling. The photograph is the size of a postage stamp. Courtesy Mr. John Waters Gatling, Asheville, North Carolina.

Richard Henry Gatling, son of Richard and Jemima Gatling. This is a tin photograph that is inscribed on the back, "1899-To My Mother." Courtesy Mr. John Waters Gatling, Asheville, North Carolina.

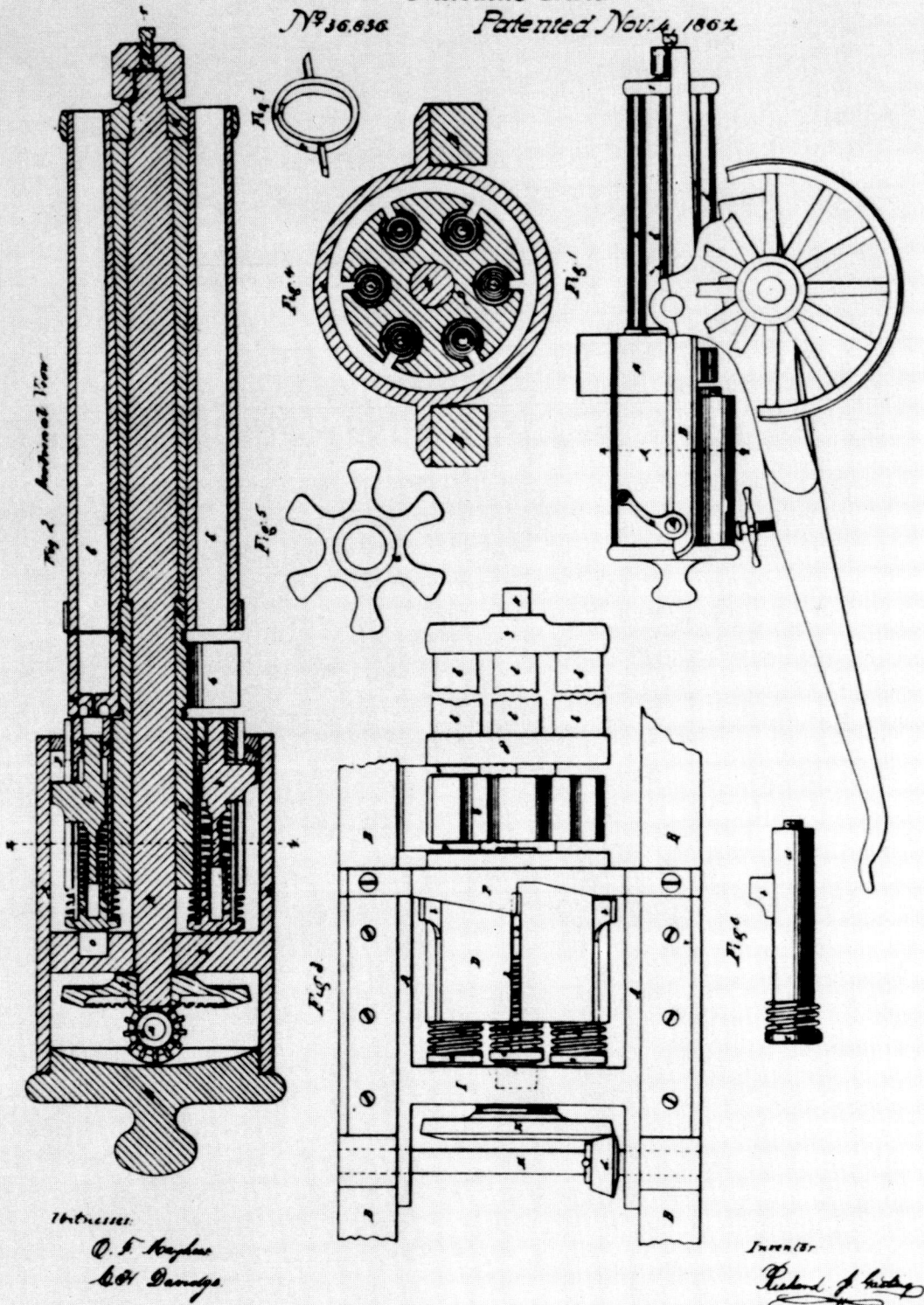

Drawing for Richard Jordan Gatling's machine gun, patented November 4, 1862. Courtesy National Archives, Washington, D.C.

Office of

Gatling Gun Company.

Hartford, Conn. U.S.A.

April 17th 1875.

My Dear Cousin

Letterhead for the Gatling Gun Company. (E. Frank Stephenson, Jr.)

Drawing of the Colt Company, Hartford, Connecticut where Richard Jordan Gatling had his guns manufactured. Courtesy Mr. John Waters Gatling, Asheville, North Carolina.

Advertisement for the Gatling Gun before Richard Jordan Gatling moved his base of operations to Hartford, Connecticut. Courtesy Mr. John Waters Gatling, Asheville, North Carolina.

44

Newly completed Gatling guns at the Colt Company in Hartford, Connecticut. Courtesy C. B. Robertson Family, Jackson, North Carolina.

Gatling Gun. Courtesy of Mr. John Waters Gatling, Asheville, North Carolina.

Gatling Gun. Courtesy of Mr. John Waters Gatling, Asheville, North Carolina.

Gatling Gun. Courtesy of Mr. John Waters Gatling, Asheville, North Carolina.

47

Gatling Gun. Courtesy of Mr. John Waters Gatling, Asheville, North Carolina.

Gatling Gun. Courtesy of Mr. John Waters Gatling, Asheville, North Carolina.

48

Gatling Gun. Courtesy of Mr. John Waters Gatling, Asheville, North Carolina.

Gatling Gun. Courtesy of Mr. John Waters Gatling, Asheville, North Carolina.

Gatling Gun. Courtesy of Mr. John Waters Gatling, Asheville, North Carolina.

Gatling Gun. Courtesy
of Mr. John Waters
Gatling, Asheville,
North Carolina.

50

Gatling Gun. Department of Defense official photograph.

Gatling gun crew on USS Alliance. Department of Defense official photograph.

51

Gatling Gun. Department of Defense official photograph.

Gatling guns in China. Courtesy Mr. John Waters Gatling, Asheville, North Carolina.

Gatling Camel Corps. Courtesy Mr. John Waters Gatling, Asheville, North Carolina.

● ●

LITTLE BIG HORN

Lt. Col. George Armstrong Custer refused to take the four Gatling guns that were at his disposal when he headed out for the Little Big Horn in June of 1876. Custer felt the Gatlings would have slowed him down and that the Gatlings were too much firepower for a "band of Indians."

Gatling Gun crew in Cuba during Spanish-American War. Courtesy Mr. John Waters Gatling, Asheville, North Carolina.

Gatling Gun crew in Philippines in 1899 in Spanish-American War. Courtesy Mr. John Waters Gatling, Asheville, North Carolina.

Gold medal awarded to Richard Jordan Gatling at the International Centennial Exposition in Philadelphia in 1876. Courtesy Mr. John Waters Gatling, Asheville, North Carolina.

Silver Medal awarded to Richard Jordan Gatling at the International Centennial Exposition in Philadelphia in 1876. Several years before the death of Richard Jordan Gatling, many of his medals and awards were stolen from his home. Courtesy Mr. John Waters Gatling, Asheville, North Carolina.

Richard Jordan Gatling. Photograph taken in Hartford, Connecticut. Courtesy Mr. John Waters Gatling, Asheville, North Carolina.

Jemima Sanders Gatling. Photograph taken in Hartford, Connecticut. Courtesy Mr. John Waters Gatling, Asheville, North Carolina.

The restored home of Mr. and Mrs. Richard Jordan Gatling, Hartford, Connecticut. Courtesy Gatling Mansion Limited Partnership comprised of Ms. Marion Chertow, Mr. Matthew Memerson, Mr. Dennis Main and Mrs. Renee Main. 1988 photograph by Ms. Marion Chertow.

Richard Jordan Gatling's daughter, Ida, in her wedding dress in the Gatling home in Hartford, Connecticut. Courtesy Miss Rebecca Gatling Long, Jackson, North Carolina.

A NOTABLE WEDDING

Marriage of Rev. H. O. Pentecost and Miss Ida Gatling – The Ceremony at the South Baptist Church – A Large and Fashionable Assemblage – The Reception at the Bride's Residence and Other Details.

The marriage of Rev. Hugh O. Pentecost, pastor of the South Baptist church, and Miss Ida, only daughter of Dr. and Mrs. R. J. Gatling, was celebrated at the South Baptist church this afternoon with imposing ceremonies. Ever since the cards of invitation were issued Hartford soceity has been awaiting this wedding as one of the most fashionable events of the season. Both the bride and groom are well known, the latter as one of our most successful ministers and the former as a general favorite. At 5 o'clock, the hour of the ceremony, the church was filled to overflowing by the hosts of friends of contracting parties. The large auditorium was richly and beautifully trimmed by a profusion of flowers, arranged masses and in exquisite designs. The pulpit platform was built out even with the front row of pews and was covered with floral decorations and tributes, while on the wall in the rear of the preacher's desk, three heavy banks of flowers sent their fragrance throughout the church. During the ceremony the bride and groom stood beneath a four post doubly decorated floral arch surmounted by a horseshoe and above all the monogram "P.G." of flowers interwoven. From the center of this magnificent design depended a massive wedding bell. The entire work was a masterpiece and was in full accord with the impressive ceremony. The officiating clergyman was Rev. George F. Pentecost, the evangelist, and brother of the groom. Rev. Dr. A. J. Sage of the First Baptist and Rev. Dr. George M. Stone of the Asylum Avenue Baptist churches were present on the platform with the officiating minister, but took no active part in the ceremonies. The bridesmaids were six in number and were Miss Daisy McCoy of New York, Miss Anna Wilson, daughter of Hon. Judge Wilson of Washington; Miss Lillie Fitzgerald of Hartford, these were dressed in pale blue satin; Miss Mary E. Robinson, niece of Hon. Henry C. Robinson, of Hartford, these were dressed in delicate pink satin. The groomsmen who previous to the ceremony officiated as ushers, were Mr. Tom S. Beaty, of New York; Mr. Joseph T. Bowen, of Hartford; Mr. John A. Welles, son of the late Hon. Gideon Welles, of Hartford; Mr. T. Belknan Beach, of Hartford; Dr. S. B. St. John, of Hartford, and Mr. Louis Q. Jones, of Hartford.

At 5 o'clock precisely the bridal procession, preceded by six groomsmen, entered the church and passing up the center aisle took their stations on the pulpit platform. The bride walked between her father and mother, leaning on the former's arm. She was elegantly and richly dressed in a heavy white satin costume, of the Princess style, with a sweeping court train of white satin brocade, figured with large crushed roses and rose buds. This elaborate dress was trimmed with Duchess lace and passamenterie and a deep fringe of pearls. In the long tulle veil were sprays of orange blossoms and cape jessamines. The diamonds worn by the bride accorded well with her costume. In her ears were a resplendent pair of solitaire earrings, gift of the groom. At her throat was a costly bar pin and in her hair a spray of the same stones flashed. To complete the costume was a white satin fan with Duchess lace cover and handkerchief to match. She also carried a bunch of loosely arranged Marshal Neil rose buds.

The bride was met at the pulpit by the groom, who awaited her there, and after the bridal party had taken their stations the services were at once begun. These were brief but very beautiful, and the large assemblage was hushed while the pair were being joined. After the ceremony the bride and groom left the church, followed by the bridesmaids and groomsmen in couples.

The music at the church was very effective. Miss Ella Fiske, the church organist, presided over the instrument, and upon the entrance of the bride rendered the exquisite Swedish wedding march. During the ceremony low music was played, and at its conclusion Mendelssohn's grand march was given.

No reception will be given at the bride's residence this evening except to the bridesmaids, their parents, the groomsmen and a few personal friends of Mr. and Mrs. Pentecost. The spacious house has been richly decorated in honor of the occasion. The bride and groom will receive the congratulations of their friends beneath a floral arch surmounted by a bow and arrow. Banks of flowers have been placed on all the mantels and mirror cases, and in front of a large mirror in the reception room a massive horse shoe of beautiful flowers has been placed. The chandeliers and other available points have been taken advantage of, and large baskets have been placed in each room. Flowers abound all through the mansion.

Among the rare and costly presents to the bride were a magnificent bronze and faience jardiniere, an elegant lace bar pin with six diamond pendants, a diamond spray pin of rose-leaves and buds, of elegant design, solitaire diamond ear-rings, three bronze mirrors, two massive bronze sconces, and exquisite tete-a-tete set, a rare pair of faience vases, and many others, rich and costly.

The decorations in the church were done by the young people's association, under the director of Mr. W. R. Morgan, their president. After the ceremony of the interior of the church was photographed by DeLamater.   Hartford   October 14, 1880   (Gatling Family Papers)

Rev. Hugh O. Pentecost, husband of Ida Gatling Pentecost. Photograph taken 1906 in New York City. Courtesy Mr. John Waters Gatling, Asheville, North Carolina.

Ida Gatling Pentecost. Photograph taken 1906 in New York City. Courtesy Mr. John Waters Gatling, Asheville, North Carolina.

Gatling Guns. Courtesy of Mr. John Waters Gatling, Asheville, North Carolina.

Richard Jordan Gatling with Model 1893 Bulldog version of his gun. This is one of the last photographs of Richard Jordan Gatling taken in Hartford, Connecticut. Courtesy Mr. John Waters Gatling, Asheville, North Carolina.

64

R. J. GATLING.
BICYCLE.

No. 519,384.   Patented May 8, 1894.

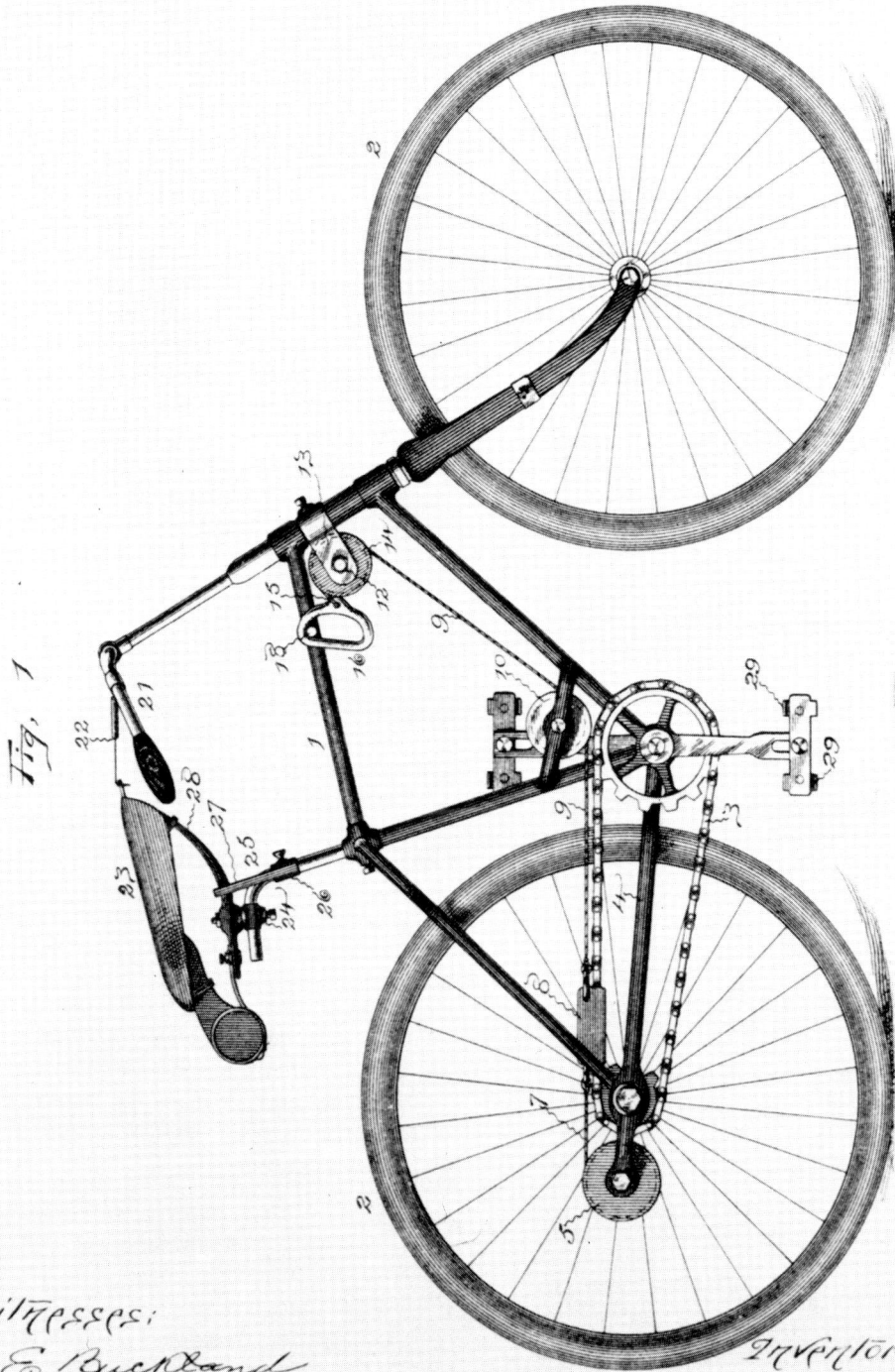

Fig. 1

Witnesses:
C. E. Buckland.
Arthur P. Day.

Inventor:
Richard J. Gatling,
by
Harry R. Williams
atty.

Drawings for Richard Jordan Gatling's bicycle, patented May 8, 1894. Drawing courtesy National Archives, Washington, D.C.

No. 668,853.

R. J. GATLING.

Patented Feb. 26, 1901.

FLUSHING APPARATUS FOR WATER CLOSETS.

(Application filed Aug. 4, 1899.)

(No Model.)

2 Sheets—Sheet 2.

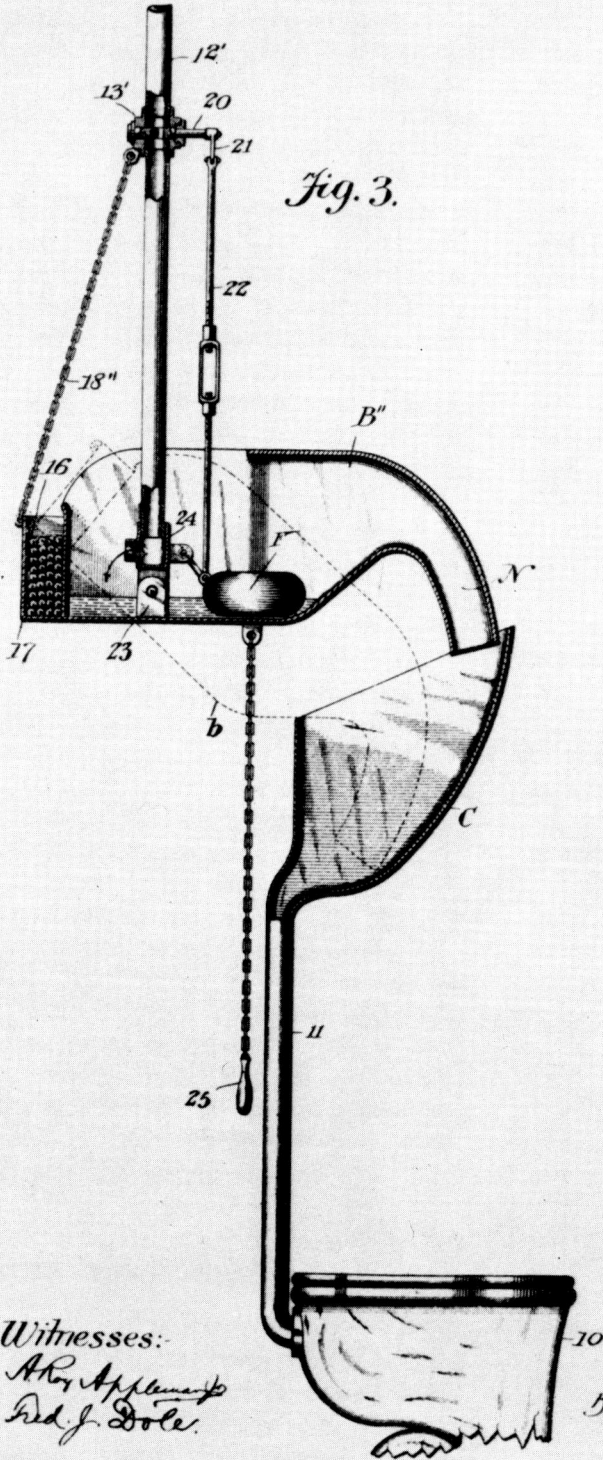

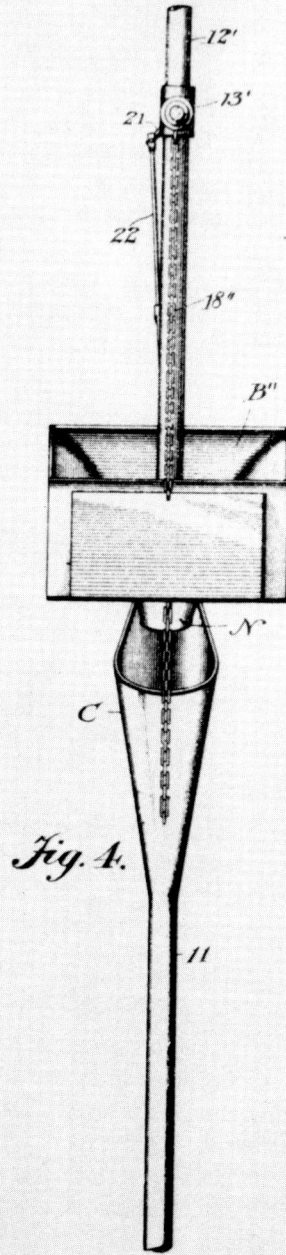

Fig. 3.

Fig. 4.

Witnesses:

A. Roy Appleman
Fred J. Dole.

Inventor
R. J. Gatling,
By his Attorney,
F. H. Richards.

Drawings for Richard Jordan Gatling's flushing toilet, patented February 26, 1901. Drawings courtesy National Archives, Washington, D.C.

66

1898 photograph of Richard Jordan Gatling and his 8 inch cast-steel gun which Congress had granted him $40,000 to help develop. Gatling had invested over $300,000 of his own money in the project which failed primarily because Gatling became too ill to supervise the test firings of the gun. Courtesy Mr. John Waters Gatling, Asheville, North Carolina.

Richard Gatling's Tractor

Roanoke-Chowan Times, Rich Square, N.C., July 31, 1902

DR. GATLING'S AUTO-PLOW - Has Invented a Gasoline Implement for Turning the soil (St. Louis Republic).

"From plowing to automobiling seems a far cry, yet those two extremes are combined in the latest inventions of Doctor R. J. Gatling, originator of the rapid firing gun which bears his name.

At the age of 80, Doctor Gatling has conceived of the idea of replacing farm horses with gasoline and changing the adornment of farmers' hands from calluses to chauffeurs' gloves.

In other words, plowing is to be revolutionized, as was modern warfare.

Many years ago the crafle took the place of the sickle, and that was later driven out of the field by the reaper, which, after a short, but useful career, was replaced by the self binding harvester. Each in its newer and better methods, cheapening the cost of producing wheat. During all this time, while the methods of harvesting the crop were being so much bettered by introducing labor saving machinery, very little progress has been made towards cheapening the cost of preparing the land for the seed.

It has remained for Doctor Gatling to invent a Motor-Plow, driven by a gasoline engine of sufficient power or propel the plows at any desired depth between one and twelve inches. The truck is built similar to those trucks used with traction engines, except that the steam boiler is replaced by a strong platform on which is placed the gasoline engine. It is connected with the tradition gearing by a series of wheels; to this truck is attached a set of disc plows.

With this machine it is estimated one man can plow from thirty to thirty five acres in a day. To plow this number of acres in one day the ordinary plow would require fifteen men and thirty horses.

All that is required to operate the Gatling Plow is for the farmer to sit upon the cushion seat of the truck and work the controller, which is not unlike those attached to automobiles. If he happens to be indisposed, his wife can take his place.

It is generally estimated that the cost of plowing under ordinary conditions is $1.50 per acre, and then the further preparation of the ground by harrowing and rolling it costs another 50 cents per acre. By the process of plowing with the Gatling machine the ground becomes thoroughly pulverized and the rolling is not required. Dr. Gatling is having his plow made in St. Louis and is going to form a St. Louis company to carry on its manufacture and distribution for the market. As yet his plans in the latter respect have not assumed definite shape. The sample plow is now nearing completion and will soon be ready for inspection.

When seen at his residence, No. 3650 Lindell Boulevard, he requested that nothing be published about his invention until after the model has been completed. He declined to talk about the wonderful mechanism of the plow, fearing that the publication of his statements would bring numerous inquiries, which he said he would have not time to answer just now. (E. Frank Stephenson, Jr.)

•••

"The work on my plow is now well under way; but I can't say when it will be finished, for the people here are the slowest mortals I have ever met. I am however, pushing them up each day and I hope to have my plow completed and tested before this fall's plowing season is over and I feel it will prove to be a great success. It always takes much time to perfect and get in on any great invention and I assure you it has been no small problem to make and complete the plow in all its details as a motor plow such as I am constructing and which promises to revolutionize plowing land."

Richard Jordan Gatling writing from St. Louis, Missouri on September 14, 1902 to his son Richard Henry Gatling in New York City. (E. Frank Stephenson, Jr.)

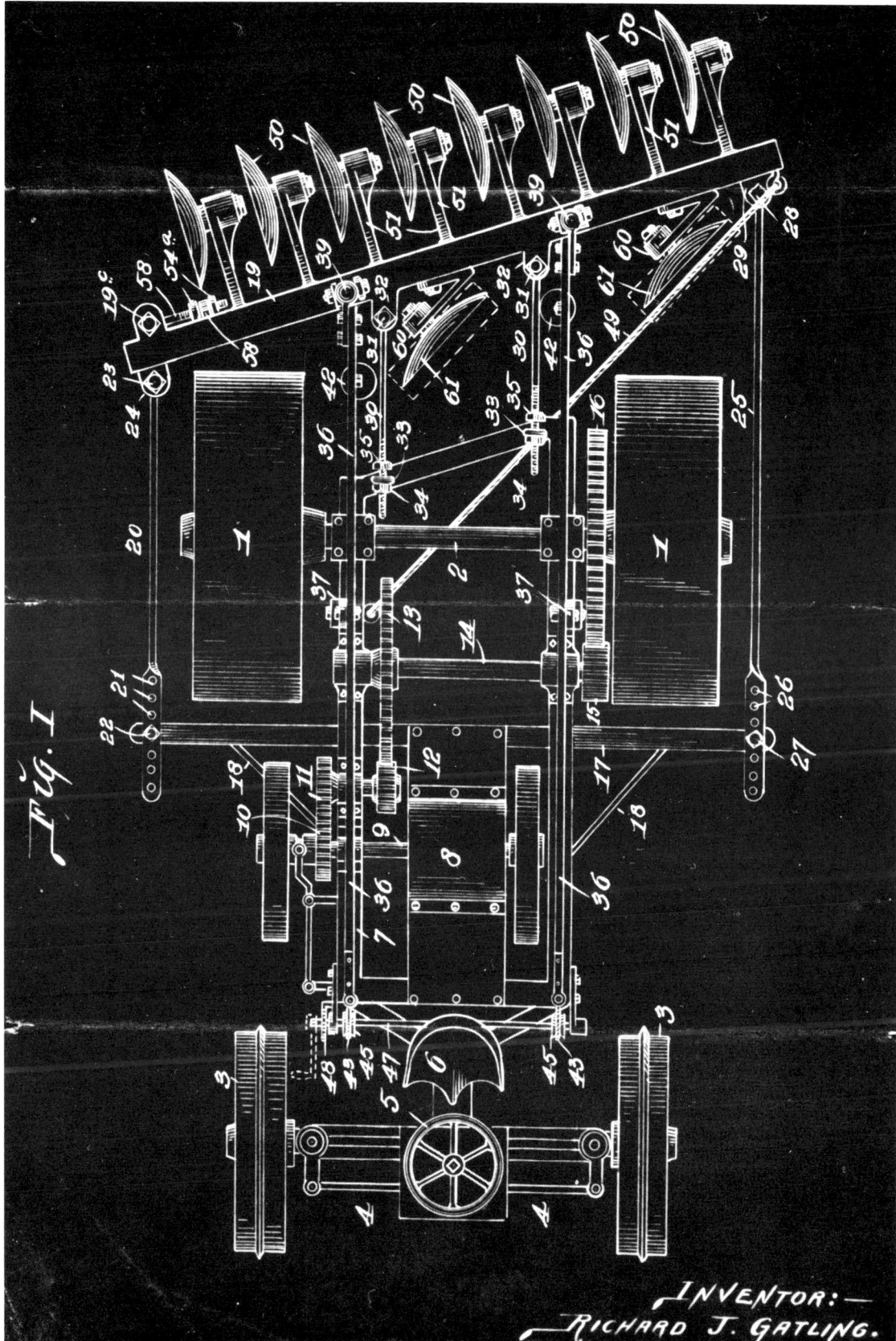

Fig. I

INVENTOR:—
RICHARD J. GATLING.

Blueprint drawings of Richard Jordan Gatling's tractor that he was working on in St. Louis in 1901-1903. Courtesy Mr. John Waters Gatling, Asheville, North Carolina.

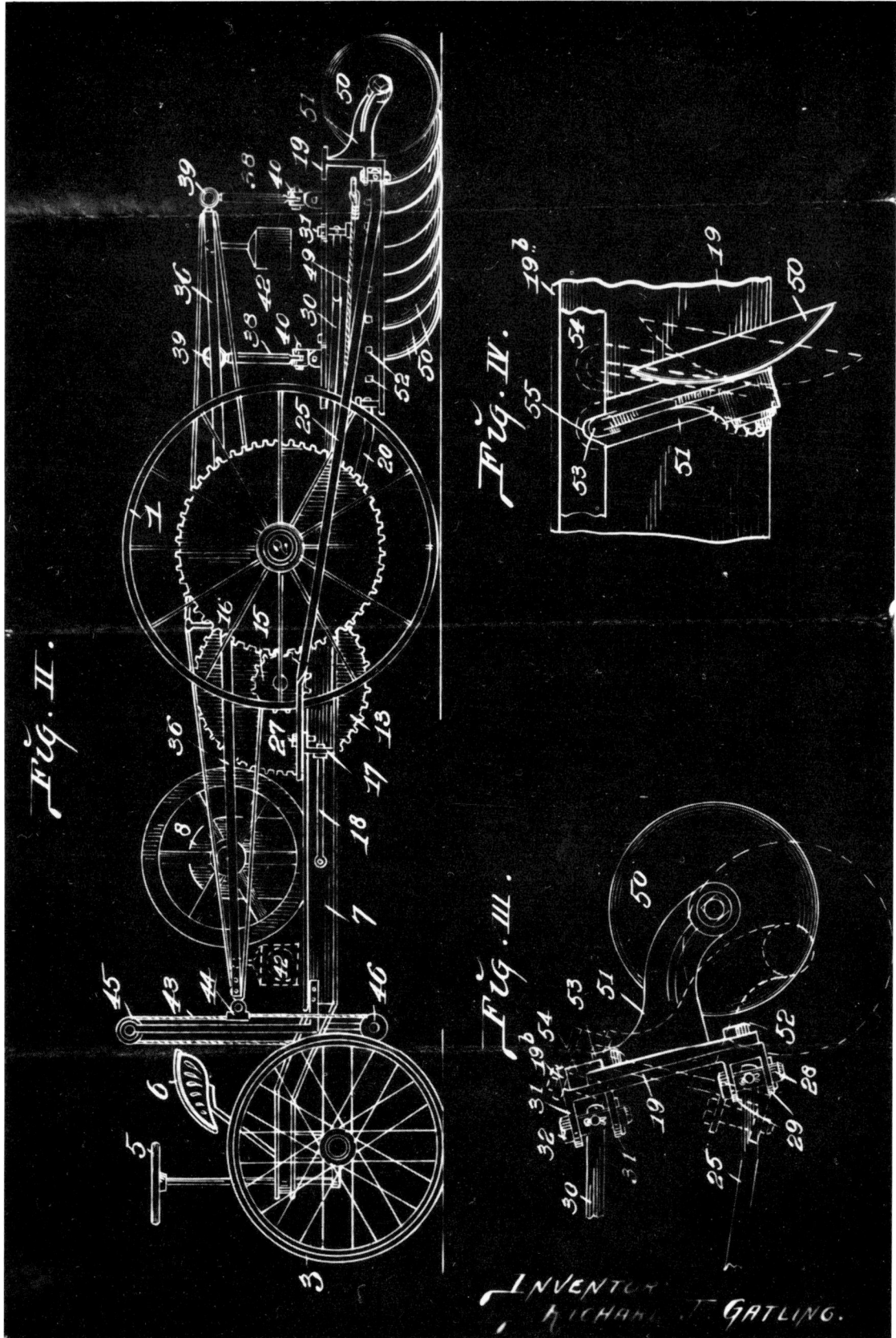

Blueprint drawings of Richard Jordan Gatling's tractor that he was working on in St. Louis in 1901-1903. Courtesy Mr. John Waters Gatling, Asheville, North Carolina.

1901 photograph of Richard Jordan Gatling with his son Richard Henry Gatling and grandson Addison Barnes Gatling. Photograph taken in New York City. Courtesy Mr. John Waters Gatling, Asheville, North Carolina.

This photograph is perhaps the last known photograph of Richard Jordan Gatling before his death in February 1903. This photograph of Mr. and Mrs. Richard Jordan Gatling was taken on December 31, 1902 at a New Year's Eve party in St. Louis, Missouri. When Richard Gatling died in 1903, he was virtually penniless, the victim of unwise investments, swindlers, overspending by family members and stiff competition from other machine gun manufacturers. Courtesy Mr. John Waters Gatling, Asheville, North Carolina.

ROANOKE CHOWAN TIMES, RICH SQUARE, NORTH CAROLINA, FEBRUARY 28, 1903
AUTHOR OF GATLING GUN DEAD
INVENTED THE WEAPON TO MAKE WAR IMPOSSIBLE.

New York, Feb. 26 – Dr. Richard Jordan Gatling, died suddenly this afternoon at the home of his daughter, Mrs. Hugh O. Pentecost, at No. 24 West One Hundred and Seventh Street. He had been downtown in the morning, attending to a business matter, and, as was his custom retired for rest after luncheon.

While his daughter was answering a telephone call from her husband, she had heard her father's labored breathing. With her mother she hurried to his couch, and, seeing that he was fast becoming unconscious, she sent for a physician.

Dr. Charles P. Duffy, who responded, administered strychnine to stimulate the action of the heart but without avail, and Dr. Gatling died in his daughter's arms within a few minutes.

Although his name has been made famous by the death-dealing Gatling gun, Dr. Gatling was a peace-loving citizen, and often declared that his one purpose in inventing the weapon was to make war impossible.

LEFT ST. LOUIS RECENTLY

Until two weeks ago Dr. Gatling was in St. Louis. The company controlling his new motor plow had its headquarters in St. Louis, under the name of the Gatling Motor Power Co., and it has been capitalized for $500,000.

For three years Dr. Gatling had been a victim of heart trouble and recently he'd had the grip. One of his sons living in New York went to St. Louis to bring him here.

At the time of his death Dr. Gatling was perfecting a few business formalities prior to placing his new motor plow on the market.

PROLIFIC INVENTOR

He had invented many agricultural devices – drills, plows, both steam and automatic before he was led to devise a military weapon, and he often said, "I would have made much more money if I had devoted myself to peace machinery instead of weapons of offense and defense.

He was born in Hertford County, N.C., Sept. 12, 1818. His genius was inherited, for his father was an inventor, too, and devised several machines used in cotton growing and mfg.

After a primary school education, young Gatling went to work with his father, but books always had a great attraction for him, and to his last days he loved history and fiction. So, at 19, it was natural to find him teaching a country school and studying himself.

ONCE A ST. LOUIS CLERK.

He went to St. Louis the following year, became a clerk in a store, and soon started in on his own account. With a little money amassed, the old trend of mind returned and he soon invented a seed-sowing machine, which almost instantly caught the fancy of the farmers of the west. (This account seems erroneous; he invented the seed sower before leaving Hertford County but adapted it to wheat drilling in St. Louis). A double-acting hemp breaker was his chief invention up to that time, but in 1857 he invented the steam plow.

The sight of the returning wounded soldiers early in the Civil War led Dr. Gatling to wonder if war could not be made so terrible that nations would hesitate to resort to it.

The application of the theory of feeding bullets as seeds are fed out of his machines occurred to him. He made a study of the principles of projectiles and evolved the idea of a sheaf of breech-loading gun barrels fed by a strip of cartridges, fired in rapid succession.

Firing from 200 to 600 shots a minute with remarkable accuracy for ranges up to 2,000 yards, the guns revoluntionized small arms theories. When their efficiency was increased so that a Gatling gun would fire 1,200 shots a minute, without any perceptible recoil and permitting either a lateral sweep while firing or the sending of every one of the shots at a small target, the complete success of the Gatling gun was assured the world over.

In 1854 Dr. Gatling married the youngest daughter of Dr. John H. Sanders of Indianapolis. St. Louis Republic.

Inventor of Famous
Gatling Gun Dead

New York Evening Telegram
Feb 26th
1903.

DOCTOR RICHARD JORDAN GATLING.
Inventor of the Famous Gatling Gun, Used in All Wars of Recent Years.

New York Evening Telegram, February 26, 1903, notice of the death of Richard Jordan Gatling. Gatling Family Papers, courtesy of Mr. John Waters Gatling, Asheville, North Carolina.

Indianapolis, Indiana, Journal, February 27, 1903 notice of the death of Richard Jordan Gatling. Gatling Family Papers, courtesy of Mr. John Waters Gatling, Asheville, North Carolina.

Indianapolis Journal

Feb 27th 1903

DEATH OF DR. GATLING

FORMER INDIANAPOLITAN WHO ACHIEVED WORLD-WIDE FAME.

Inventor of the Gatling Gun, Grain Drill and Other Devices Which Have Benefited Many.

BRIEF SKETCH OF HIS CAREER

STUDIED MEDICINE AND WORKED ON INVENTIONS WHILE HERE.

Reminiscences of His Visit to Europe —His Impressions of Crowned Heads—Married in This City.

NEW YORK, Feb. 26.—Richard J. Gatling, inventor of the Gatling gun, died here to-day at the home of his son-in-law, Hugh O. Pentecost. At 1:15 o'clock Dr. Gatling returned to the home of his son-in-law from a trip down town on business at the offices of the Scientific American. Being eighty-four years old, and accustomed to resting after any physical effort, he told his daughter, Mrs. Pentecost, he would lie down. Shortly afterwards he commenced to breathe heavily and a physician was called, who administered strychnine, but to no avail, and Dr. Gatling died in his daughter's arms a few minutes afterwards.

Richard Jordan Gatling was born in Hartford county, North Carolina, Sept. 12, 1818. While yet a boy he assisted his father in perfecting a machine for sowing cotton seed

New York Herald

Feb. 27-1903.

HERALD, FRIDAY, FEBRUARY 27, 1903.—TWE

DR. RICHARD J. GATLING.
FAMOUS GUN INVENTOR WHO DIED YESTERDAY.

DR. GATLING DIES SUDDENLY AT HOME

Worldwide His Fame as Inventor of the Gun Bearing His Name.

DR. GATLING DEAD.

CELEBRATED INVENTOR OF THE GATLING GUN.

AN AMERICAN GENIUS OF MARKED INVENTIVE ABILITY.

Formerly President of the Gatling Gun Company and a Resident of This City — Removed About Ten Years Ago.

Dr. Richard Jordan Gatling, inventor of the world-famous Gatling machine gun, died suddenly yesterday afternoon at his home in New York, the residence of his daughter, Mrs. Hugh O. Pentecost. He returned about 1:15 o'clock from a trip down town to the office of the "Scientific American." Being accustomed to resting after any physical effort, he told his daughter that he would lie down for a while. Soon after he began to breathe heavily and a physician was called who administered strychnine, but to no avail, and the doctor died in his daughter's arms in a few moments.

Richard Jordan Gatling, the inventor of the celebrated revolving battery gun which bears his name, was born in Hertford county, N. C., September 12, 1818. His father was a substintial, industrious farmer, a man of great energy of character. Although a slave-holder, he taught his children the necessity of labor and economy as the surest road to fortune. When but a lad, the subject of this sketch assisted his father in the invention of a machine for sowing cotton; also a ma-

drive small engines by compressed air, instead of by steam. By such a distribution of power, all small furnaces and coal deposits used for driving small steam engines, he thought, could be dispensed with.

About 1850, he invented a double-acting hemp-brake, which is still used in some parts of the West, for breaking hemp. In the year 1857, he invented a steam plow, or earth-pulverizing machine, designed to be propelled or operated by combined animal and steam power; but ill-health, and the low price of grain in the West at the time, prevented him from working out the details of this machine to practical results.

In 1861 he conceived the idea of making a machine gun, which would, to a great extent, supersede the necessity of large armies. He made his first revolving battery gun while living in Indianapolis, and in the spring of 1862 he fired it, in its then imperfect state, over 200 shots per minute, in the presence of many army officers and private citizens. In the fall of the same year, he went to the city of Cincinnati, O., and had a battery of six of his guns made at the manufacturing establishment of Miles Greenwood & Co.; but about the time the guns were completed the factory was burned, with the guns, subjecting him to heavy pecuniary loss. He then had twelve of his guns made at another establishment in Cincinnati, which were afterward used by General Butler at Petersburg. In 1865, he made additional improvements in his battery gun. test trials of which were made at Washington the same year; and in 1866, he had his guns made at the Coopers' Firearms Manufacturing Company, Philadelphia, and thorough test trials were made with it at the Frankfort Arsenal, Philadelphia, and subsequent trials were also made at Washington and Fortress Monroe, which proved to be so satisfactory as to induce Mr. Stanton, then secretary of war, and General A. B. Dyer, chief of ordnance, to adopt the arm into the service. In August, 1866, an order was given for 100 of these weapons, fifty of one-inch and fifty of .50 of an inch caliber, which arms were made at Colt's Armory and delivered in 1867.

In 1867 Dr. Gatling went to Europe and spent eighteen months in presenting the merits of his gun to foreign governments. He made a second trip three years later and upon his return settled in this city. Ten years later he visited England again. Since the approval of the Gatling gun by the United

Hartford, Connecticut, _Courant_, February 27, 1903 notice of the death of Richard Jordan Gatling. Gatling Family Papers, courtesy of Mr. John Waters Gatling, Asheville, North Carolina.

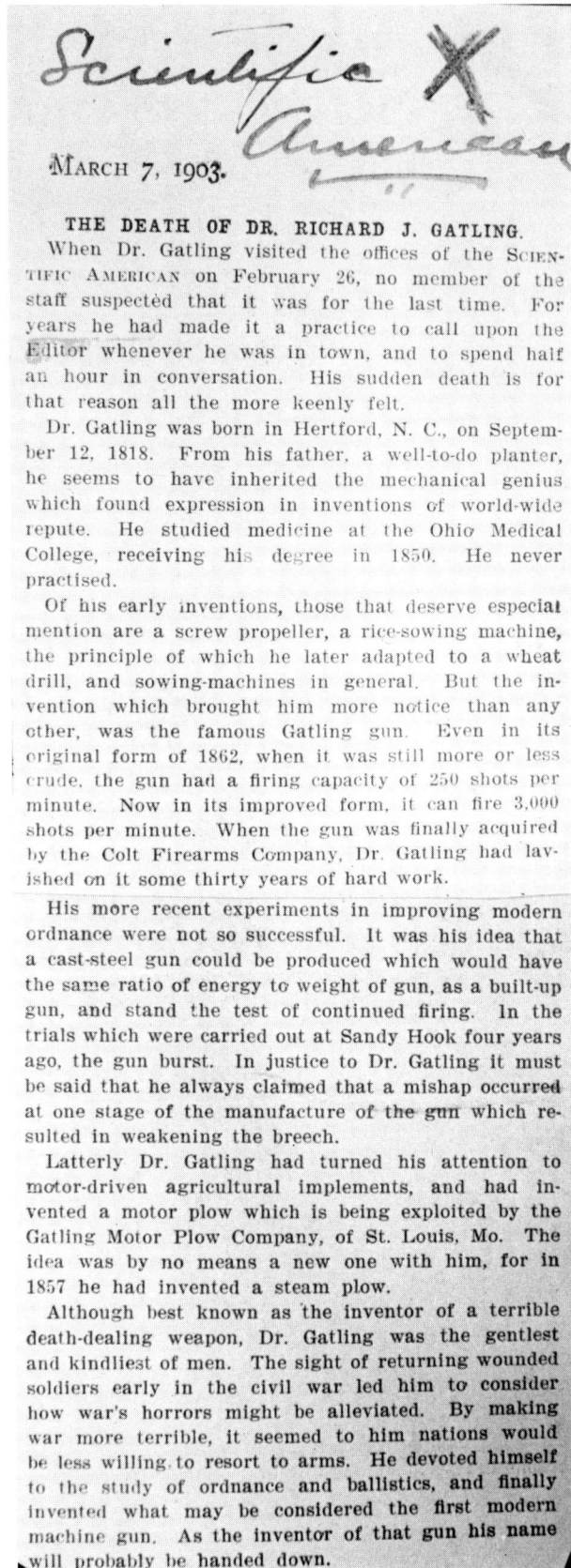

Scientific American

MARCH 7, 1903.

THE DEATH OF DR. RICHARD J. GATLING.

When Dr. Gatling visited the offices of the SCIENTIFIC AMERICAN on February 26, no member of the staff suspected that it was for the last time. For years he had made it a practice to call upon the Editor whenever he was in town, and to spend half an hour in conversation. His sudden death is for that reason all the more keenly felt.

Dr. Gatling was born in Hertford, N. C., on September 12, 1818. From his father, a well-to-do planter, he seems to have inherited the mechanical genius which found expression in inventions of world-wide repute. He studied medicine at the Ohio Medical College, receiving his degree in 1850. He never practised.

Of his early inventions, those that deserve especial mention are a screw propeller, a rice-sowing machine, the principle of which he later adapted to a wheat drill, and sowing-machines in general. But the invention which brought him more notice than any other, was the famous Gatling gun. Even in its original form of 1862, when it was still more or less crude, the gun had a firing capacity of 250 shots per minute. Now in its improved form, it can fire 3,000 shots per minute. When the gun was finally acquired by the Colt Firearms Company, Dr. Gatling had lavished on it some thirty years of hard work.

His more recent experiments in improving modern ordnance were not so successful. It was his idea that a cast-steel gun could be produced which would have the same ratio of energy to weight of gun, as a built-up gun, and stand the test of continued firing. In the trials which were carried out at Sandy Hook four years ago, the gun burst. In justice to Dr. Gatling it must be said that he always claimed that a mishap occurred at one stage of the manufacture of the gun which resulted in weakening the breech.

Latterly Dr. Gatling had turned his attention to motor-driven agricultural implements, and had invented a motor plow which is being exploited by the Gatling Motor Plow Company, of St. Louis, Mo. The idea was by no means a new one with him, for in 1857 he had invented a steam plow.

Although best known as the inventor of a terrible death-dealing weapon, Dr. Gatling was the gentlest and kindliest of men. The sight of returning wounded soldiers early in the civil war led him to consider how war's horrors might be alleviated. By making war more terrible, it seemed to him nations would be less willing to resort to arms. He devoted himself to the study of ordnance and ballistics, and finally invented what may be considered the first modern machine gun. As the inventor of that gun his name will probably be handed down.

78

The death of Dr. Gatling removes a figure of national importance in the early days of the war and of world-wide reputation. In the days before the war, as a resident of Indianapolis, few were better known here than he. Connected by marriage with one of the oldest families that were among those that founded Indianapolis, he had long been a prominent figure, and when he had invented the gatling gun — as it became known throughtout the world — he attained a celebrity that added to the general regard in which he was held. He invented his gun here, and here most of the early work was done on it and the project placed on its feet. From his early days he had been absorbed with problems after the manner of the born inventor, and his triumph was a gratification to the wide circle that knew and valued him for his gentle character and high worth. His venerable age had carried him past the day of those that were his contemporaries in his great work and the latter half of his life he lived in the East. But Indianapolis never forgot the kindly gentleman whose mysterious pursuits finally crystallized in his great invention. He did his part in life well — both to labor and to wait — and he leaves a memory fragrant with affectionate remembrance.

Undated editorial from an Indianapolis, Indiana newspaper. Gatling Family Papers, courtesy of Mr. John Waters Gatling, Asheville, North Carolina.

Hartford Courant — DR. GATLING Editorial — Feb 27-1903.

Any catalogue of the Hartford names that have gone round the world would include his. Hartford did not lose her right and title in him when—already an aged man—he ceased to walk her streets. The work that made him famous was wrought here; he remains for all time one of the town's assets.

It is a rather curious circumstance that two such Hartforders as General Hawley and Dr. Gatling should have been born in North Carolina. In the quality and bent of his mind the doctor was from the first a Connecticut Yankee. He was "projickin' " with wheels and cogs and such like mechanical mysteries before he was out of his teens. Before he was 25 he was inventing things and applying for patents. Some of those earlier inventions of his had merit in them, too, and money. They were also valuable practice for him—a preparation.

The year 1861 came and the war with it—a tremendous stimulant to talent and to genius. Presently people were talking about the Gatling gun—the "coffee-mill" the old soldiers called it. Reports of the wonderful way in which it ground out death and wounds spread through the land and crossed the sea. There are other machine-guns now, but that was the first of the family. It made the doctor a famous man.

His old friends here have often been tickled by the whimsical, humorous incongruousness of it all. For Dr. Gatling was not at all the kind of man you might expect a gun-inventor to be—a war-loving man, a man of blood. He was an intellectual, amiable gentleman, fond of his books, his workshop, his club, his game of cards, a good story, a jolly company. He was himself companionable, an entertaining talker, a capital story-teller. But that invention of his has not said its last word to the world yet. Napoleon's "whiff of grape shot" is historic, and Napoleon did not have the machine-gun.

We see by a comparison of dates that next September would have brought Dr. Gatling's eighty-fifth birthday.

Dr. Gatling.

To the Editor of The Courant:—

Dr. Gatling's family are grateful to you for the appreciative editorial you published concerning him. You caught the spirit of the man with accuracy and sympathy.

Dr. Gatling was an exception to the rule that no man is great to his valet. His distinguishing characteristics were fearlessness, indomitable perseverance and gentleness. Sweetness and gentleness usually go with timidity and weakness of purpose, but he was at once the strongest and tenderest of men, and he has left the impression upon those of us who knew him in his home relations that we had with us a truly great man.

Only a few days before his death he told me that to him the whole value of life was in that it was a preparation for a happy immortality, and that above all his earthly ambitions was his des   to be spiritually prepared for im   rtality. This habitual state of his mi d and the manner of his going robbed death of every terror and made it se m a thing of beauty.

Hugh O. Pentecost.
249 West 107th street.
New York, Feb. 28, 1903.

Gatling family's response to the Hartford, <u>Courant</u> editorial. This letter, written by Richard Jordan Gatling's son-in-law, Hugh O. Pentecost appeared in the March 2, 1903 issue of the Hartford <u>Courant</u>.

Indianapolis Sentinel
March 2nd
1903.

DR. GATLING'S FUNERAL

SERVICES THIS AFTERNOON
FROM J. R. WILSON'S HOME.

Interesting Career of the Noted Invent-
or—Famous Inventions the Prod-
uct of His Fertile Brain.

The remains of Dr. Richard J. Gatling
will arrive in Indianapolis this morning
over the Big Four at 11:45 o'clock. They
will be met at the Union station by
friends and relatives of the deceased and
taken to the residence of John R. Wilson,

Indianapolis, Indiana, Sentinel, March 2, 1903 notice of the funeral of Richard Jordan Gatling.
Gatling Family Papers, courtesy of Mr. John Waters Gatling, Asheville, North Carolina.

82

ADDRESS DELIVERED BY THE REV. A. R. BENTON AT THE FUNERAL OF DR. RICHARD J. GATLING, INDIANAPOLIS, IND., MARCH 2nd, 1903.

We meet in the solemn stillness of this home to honor, as we may, one who has been greatly honored both at home and abroad. We also gather here to express our deep sympathy and tender regard for the family, so sadly bereft of husband, father, wise counselor and loving companion.

It seems fitting that these last sad offices due to our departed friend, should be performed in this our commonwealth, in this City and this home. For many years of his early life, his residence was here; here his family relations were made, and here his life's work was successfully begun. Here he became notable among the men of this commonwealth.

Among the men of note, who have made the state famous, who, as statesmen, authors, men of affairs have won for our state a high place in the world of progress, culture and civilization, will be enrolled the name of Richard J. Gatling.

It is also fitting that he find his last resting among his relatives and loved ones who sleep in the beautiful City of our dead.

An acquaintance with our departed friend of nearly fifty years, justifies me in speaking with freedom of the personal traits and public service of him, whose loss we mourn to-day.

The greatest of Roman historians, writing of his departed friend, said of him - "He was truly happy in the distinctions of his life and happy in the occasion of his death." These words give admirable expression to the achievements and death of our friend, and in a broader and truer sense are applicable to the latter rather than to the former.

In every life, its character is determined by two essential factors. These are inherited traits and tendencies and the environment of the time, and the spirit of the passing age. To both of these influences every life must respond in a large measure, and from them comes the call for one's life work. Thus considered, a life, when developed naturally, is not a haphazard, inconsequential growth, but is orderly and along the line of natural endowments and surrounding conditions.

Dr. Gatling, as I knew him, possessed a temperament and mental traits of a high order, that made possible the achievements of his active career. His intellectual grasp of subjects was strong, his philosophical tendency was marked, and his constructive imagination made him an inventor. These qualities determined his career which he so grandly pushed to final success.

But, to his intimate friends, the most attractive qualities he possessed were found in his gentle spirit, and unfailing kindness toward others, as husband, father, friend. There was a "sweet reasonableness" as a golden thread, running through his whole life, making him tolerant of others, splendidly courteous in intercourse and kindly considerate of the rights and sentiments of others. To these high qualities of refined feeling we all can bear witness and remembering him, the ideal of a true gentleman will never fade from our minds.

Co-ordinated with this gently spirit and kindliness, there was a firm persistent purpose to accomplish to task to which he had been called and to which he set himself with unfaltering devotion. This task, as events proved, called out the most heroic efforts and untiring energy, all of which were met with an optimistic faith in his final success.

The life of our friend was cast in a time of great events. It was the storm and stress period of our national history. Men's minds were turned toward the settlement of the great West and were stirred by the events of the Civil War. These environments shaped the productive labors of Dr. Gatling.

Interested in the toils of the husband-men, in order to lighten labor and to facilitate production, to the end that the vast, unsettled tracts lying west of the Mississippi might be subdued, he first devoted his inventive genius to the construction of implements for the farmer. Soon came the stress of our Civil War and he conceived the idea of mitigating the horrors of war by a more deadly weapon than any one then in use. Paradoxical as it may seem, it was his contention that the more destruction the weapons of war, the fewer are its casualities. In his earnest and convincing way he maintained, his aim to invent such a weapon, was humane, beneficent and philanthropic. Nor was his contention without reason, for he had all history behind him, supporting its claim. This destructive arm of warfare, now bearing his name, has become famous on two continents, and promises to perpetuate his name and fame to remote generations.

He was also happy in the occasion of his death. In his gentle way he said he would take a little rest. Reclining, he closed his eyes to look no more on this fair world. He opened them on fairer scences, and brighter visions of the spirit land. In this restful, painless, quiet passing to the great beyond, how happy his lot. In this time world he was happy in a long life, in the esteem of troops of friends, in a most honorable and useful career; but thrice happy in the time and manner of his death and in the greetings of loved ones gone before.

As he closed his eyes for a moment of rest we may imagine him in soliloquy indulging in this apostrophe to his life, beautiful as it is applicable:

> Life! we've been long together,
>   Through pleasant and through cloudy weather
> Tis hard to part when friends are dear,
>   Perhaps 'twill cast a sigh, a tear.
> Then steal away, give little warning
>   Choose thine own time; say not good night.
> But in some brighter clime
>   Bid me good morning.

And he stole away from us without warning, to be greeted in that brighter clime with a word of cheerful welcome, and its good morning.

He being dead yet speaketh. He speaks inner memories of his kindly spirit, his deeds of goodness, and of his words of cheerfulness and hope. These voices of our dead shall be ever in our ears and be heard with reverent attention.

To you who have so suddenly lost such a beloved friend, companion and helper, I can only say, you have a Father in Heaven, infinite in his love and tender compassion whom you may trust, and on whom you may lean with confidence and who ever doeth all things well. And so, may the Lord bless and be gracious to you; the Lord lift upon thee the light of understanding - a peace that comes from God the living Father, and from Our Lord Jesus Christ. (Gatling Family Papers)

•••

SUGGESTED READING

Johnson, F. Roy and Stephenson, Jr., E. Frank. The Gatling Gun and Flying Machine, Johnson Publishing Company, Murfreesboro, North Carolina, 1979.

Parramore, Thomas C., "The North Carolina Background of Richard Jordan Gatling," The North Carolina Historical Review, Vol. XLI, No. 1, (Jan. 1964), pp. 54-61.

Wahl, Paul, and Toppel, Donald. The Gatling Gun, Arco Publishing Company, Inc., New York, New York, 1965.

Gatling Monument in Crown Hill Cemetery, Indianapolis, Indiana. Photograph taken 1950 by a cousin of Richard Jordan Gatling, Raymond Kitchen of Como, North Carolina.

MONUMENT OF DR. RICHARD JORDAN GATLING
CROWN HILL CEMETERY INDIANAPOLIS, INDIANA

The Gatling monument is constructed of Barre, Vermont granite and was assembled by Goth and Company of Indianapolis. The monument erected about 1915 was cut and furnished to them by A. Barclay and Company of Barre, Vermont. The base is 10 feet x 5 feet x 1 foot and is of axed granite. The die is 8 feet x 1 foot x 8 feet, and it is also axed. Four Tuscan columns 1 foot x 1 foot x 8 feet rest on the base and carry part of the load of the cap. The columns are hammer finished. The cap is 8 feet 8 inches x 2 feet 11 inches x 3 feet. The total height of the monument is 12 feet. Set in the rear of the die is a panel containing the inscriptions of the Gatling children. This panel is 3 feet x 5 feet. The bronze tablet on the front side of the die is 3 feet x 5 feet and contains the inscriptions of Dr. Gatling and his wife. This tablet was designed and cast by Rudolph Schwartz, sculptor, Indianapolis, Indiana.

DR. RICHARD J. GATLING, inventor of the Gatling Gun, born Hertford County, North Carolina, September 12, 1818, died New York City, February 26, 1903. The highest honors that the world can boast are subjects far too low for my desire. The brightest dreams of glory are at most incomplete compared to my belief in the immortality of the soul.

JEMIMA TAYLOR SANDERS, his beloved and saintly wife, born May 27, 1837, died September 26, 1908. Whose 48 years of married life as wife and mother were filled with unfailing tenderness and devotion.

MARY S. GATLING, born October 5, 1855, died February 11, 1860.

IDA GATLING PENTECOST, born September 5, 1858, died January 13, 1911.

RICHARD HENRY GATLING, born March 7, 1870, died January 14, 1941.

ROBERT B. GATLING, born October 9, 1872, died December 2, 1903.

Note: The Gatling Crown Hill Cemetery plot also contains the graves of the two deceased young children of Richard and Jemima Gatling, William and Mary. There are no markers to their graves.

The Gatling's long time cook and housekeeper, Rachel Stepney, a black woman who died in 1885 is also buried here. There is a stone to her just to the right of the Gatling monument. Rachel Stepney, born 1815, was a wedding gift to Richard Gatling and Jemima Sanders by her father when they were married in Indianapolis, Indiana on October 24, 1854. Richard Gatling immediately freed Rachel Stepney but she elected to stay with the Gatlings as a free woman for the rest of her life.

●●●

ABOUT THE AUTHOR

E. Frank Stephenson, Jr., has been associated with Chowan College, Murfreesboro, North Carolina for the past twenty-five years as an administrative officer. He is a graduate of Chowan College and North Carolina State University. Among his writing credits are six books and numerous magazine and newspaper articles. Included in his upcoming books are North Carolina's Herring Fishermen, a photographic documentary on the herring fishermen on the Chowan and Meherrin rivers, and Carolina Moonshine Raiders, a look back at the moonshine busters in the Roanoke Chowan region of North Carolina in the period 1940-1970.

Gatling Monument in Jordan Gatling Cemetery, Jordan Gatling Plantation, Como, North Carolina. Photograph by E. Frank Stephenson, Jr., April 1993.

THE JORDAN GATLING CEMETERY
JORDAN GATLING PLANTATION
MANEY'S NECK TOWNSHIP
HERTFORD COUNTY, NORTH CAROLINA

In 1875 Charles Henry Foster, writing in Potter's <u>American Monthly</u>, described the Gatling burying-ground as "the handsomest private cemetery in the county [of Hertford]. On rising ground, it was enclosed by an elegant iron rail fence, octagonal in form, each of the sides being 22 1/2 feet long. The fence was erected in 1861 and was cast in Norfolk, Virginia by S. H. Hodges and Company. Just within the enclosure is a circle of box extending around its entire circumference, with an inner circle of the same perennial shrub. Hollies, junipers, and arbor vitae are placed at intervals in regular order. The whole is kept clear of grass and weeds. In the centre is a monument on a granite pedestal with a marble base, erected A.D. 1860. It is 14 feet in height. The obelisk is of white Italian marble, with delicate blue veins. It has on one face a floral wreath encircling the name "Gatling," and surmounted by the sculptured vase of flowers. On a lower corner of the monument appears the stone cutter's mark 'T. M'Caffrey Norfolk, Va.' James Henry Gatling had the monument and fence erected in honor of his parents. The graves run east and west; those of the parents are at the east of the monument. The inscriptions are:

Jordan Gatling, born May 14, 1783, died 64 years, 10 months and 29 days. He was a kind father.

Mary, wife of Jordan Gatling, born October 30, 1795, died September 30, 1868. A devoted wife, an affectionate mother, and a true Christian.

Thomas B. Gatling, died May 14, 1857, age 45 years, 7 months and 15 days. He was a useful man.

His wife Rebecca Jane, born October 14, 1815, died March 27, 1870.

Mary Ann Gatling, died October 4, 1838, age 25 years and 5 days. She died a Christian.

James H. Gatling, born July 15, 1816, died by the hands of an assassin September 2, 1879. An industrious, honest and worthy man.

Caroline Gatling, died October 23, 1836, aged 29 days. She died an angel.

Martha S. Gatling, died July 22, 1846, aged 17 years, 9 months and 23 days. She was much beloved.

Alic, infant daughter of Thomas B. and Jane Gatling.

The Jordan Gatling Cemetery is maintained by Mr. John W. Gatling, Asheville, North Carolina, Mr. Robert Lee Gatling, Ahoskie, North Carolina and The Murfreesboro Historical Association, Murfreesboro, North Carolina.

●●●

GATLING PATENTS

Richard Jordan Gatling had over fifty patents to his credit at the time of his death in 1903. His first, a patent for a seed sower, was granted on May 10, 1844. His last, a patent for a steam-plow or tractor, was dated July 22, 1902. Jordan Gatling, Richard's father, was granted patents in 1835 for a cotton thinner and cotton seed planter. James Henry Gatling, brother of Richard Jordan Gatling, received a patent in 1871 for the treatment of timber from old-field pines.

1955 view of Gatling Monument in Jordan Gatling Cemetery, Jordan Gatling Plantation, Como, North Carolina. Note the Jordan Gatling House in background. Photograph by F. Roy Johnson, Murfreesboro, North Carolina.

Jordan Gatling Cemetery, Jordan Gatling Plantation, Como, North Carolina. Photograph by E. Frank Stephenson, Jr., June 1993.

Details of Jordan Gatling Cemetery fence, Jordan Gatling Plantation, Como, North Crolina. Photograph by Colbert P. Howell and E. Frank Stephenson, Jr., 1970.

Details of Jordan Gatling Cemetery fence, Jordan Gatling Plantation, Como, North Carolina. Photograph by Colbert P. Howell and E. Frank Stephenson, Jr., 1970.

Details of Jordan Gatling Cemetery fence, Jordan Gatling Plantation, Como, North Carolina. Photograph by Colbert P. Howell and E. Frank Stephenson, Jr., 1970.

Details of Jordan Gatling Cemetery, Jordan Gatling Plantation, Como, North Carolina. Photograph by Colbert P. Howell and E. Fank Stephenson, Jr., 1973.

James Gatling Cemetery, Maney's Neck Township (Como), Hertford County, North Carolina. The lone trees mark the site of the cemetery where Richard Jordan Gatling's grandfather, James Gatling was buried in 1822. The cemetery which is located about three-quarters of a mile south of the village of Como, also is the final resting place of Roschus Gatling and other close relatives of Richard Jordan Gatling, including his great grandfather William Jordan Gatling. For many years the cemetery was maintained by William Thomas Bowles, grandson of Roschus Gatling. Today the cemetery is maintained by Harold Gatling, great-grandson of Roschus Gatling. Photograph by E. Frank Stephenson, Jr., June 1993.

Roschus Gatling with his wife, Martha Jane Pope Gatling and their children. Photograph taken in early 1890's. Photograph courtesy of the late William Thomas Bowles, grandson of Roschus Gatling.

1900 photograph of a group of Richard Jordan Gatling's Como, North Carolina cousins. Photograph courtesy of the late William Thomas Bowles, Como, North Carolina.

A reunion of the John Gatling family at the John Gatling home in Mapleton, North Carolina. The photograph taken about 1912 courtesy Mr. Robert Lee Gatling, Ahoskie, North Carolina.

Buckhorn Baptist Church, Como, North Carolina. The home church of the Jordan Gatling Family. Richard Jordan Gatling attended this church before he went west in 1844 and visited it on his trips back to North Carolina. Buckhorn Academy, a highly regarded 19th century classical school for boys, where Richard Jordan Gatling received his schooling, stood on the northeast corner of the church grounds for approximately eighty years. Photograph by E. Frank Stephenson, Jr., March 1993.

FEDERAL SHIPBUILDING AND DRY DOCK COMPANY

UNITED STATES STEEL ⓤⓢ CORPORATION SUBSIDIARY

KEARNY, NEW JERSEY

requests the pleasure of your presence at the launchings

in the Kearny, New Jersey, shipyard

on Sunday morning, June the twentieth,

at eleven-fifteen o'clock, (E.W.T.)

of the following two destroyers:

U. S. S. DORTCH (DD670)

Sponsored by

MISS MARY CLARE DORTCH

U. S. S. GATLING (DD671)

Sponsored by

MRS. JOHN W. GATLING

★

This invitation will admit you to the launchings.

Cameras will not be permitted.

Invitation to the launching of the destroyer Richard Jordan Gatling (DD671) on June 20, 1943. Gatling Family Papers, courtesy Mr. John Waters Gatling, Asheville, North Carolina.

Mr. and Mrs. John Waters Gatling at the launching of the USS Richard Jordan Gatling, June 20, 1943. Courtesy Mr. John Waters Gatling, Asheville, North Carolina.

Mrs. John Waters Gatling christens the USS Richard Jordan Gatling at its launching on June 20, 1943. Courtesy Mr. John Waters Gatling, Asheville, North Carolina.

Destroyer Richard Jordan Gatling going down the "ways" at its launching on June 20, 1943 in Kearny, New Jersey. The ship was sponsored and christened by Mrs. John Waters Gatling. Courtesy Mr. John Waters Gatling, Asheville, North Carolina.

Society of the Sponsors of United States Navy Ships medal/ribbon given to Mrs. John Waters Gatling as sponsor of the Destroyer USS Richard Jordan Gatling. Mrs. Gatling proudly wore the medal as a member of the society. Courtesy The Murfreesboro Historical Association, Murfreesboro, North Carolina.

Members of the Richard Jordan Gatling family that had gathered in Kearny, New Jersey on June 20, 1943 for the launching of the USS Richard Jordan Gatling. Courtesy Mr. John Waters Gatling, Asheville, North Carolina.

Destroyer Richard Jordan Gatling plaque given to Mrs. John W. Gatling by the captain of the USS Gatling. Courtesy The Murfreesboro Historical Association, Murfreesboro, North Carolina.

The Commanding Officer, Officers

and Crew of the

U. S. S. GATLING

request the pleasure of your company at the

Commissioning Ceremonies

to be held on board the

U. S. S. Gatling

August   1943 at   o'clock

Navy Yard, New York

R. S. V. P.
  Prior August 9, 1943
  U. S. S. Gatling
  Fleet Post Office
  New York, N. Y.

You must not discuss this event with others.

Invitation to the commissioning ceremonies for the USS Richard Jordan Gatling on August 14, 1943 at the New York Naval Yard. Gatling Family Papers, courtesy Mr. John Waters Gatling, Asheville, North Carolina.

USS Richard Jordan Gatling at its commissioning ceremonies on August 14, 1943 at the New York Naval Yard. Department of Defense official photograph.

USS Richard Jordan Gatling. Department of Defense official photograph. The ship served with distinction during World War II, mainly in the Pacific.

USS Richard Jordan Gatling. Ship was scraped in the mid 1970's. Department of Defense official photograph.

Silver service from the USS Richard Jordan Gatling. Courtesy The Murfreesboro Historical Association, Murfreesboro, North Carolina. Photograph by E. Frank Stephenson, Jr.

Silver service from the USS Richard Jordan Gatling. Courtesy The Murfreesboro Historical Association, Murfreesboro, North Carolina. Photograph by E. Frank Stephenson, Jr.

The Liberty Ship, Richard Jordan Gatling. The ship was built and launched on October 14, 1942 at the Permanete Shipyard, Richmond, California. Her sponsor was Miss Renee Richardson. None of the inventor's family were aware of the ship's construction at that time. Photograph taken on May 26, 1943 by the United States Coast Guard.

Machine-Gun Inventor Hoped to Aid Mankind

Letter Reveals Gatling's Plans to Obviate Large Armies

FORT WAYNE, Ind., Nov. 23 (UP).—According to Dr. L. A. Warren, director of the Lincoln Museum in Fort Wayne, the inventor of the machine gun hoped to aid mankind rather than destroy it.

In a letter owned by the museum and written by Richard Gatling on June 15, 1877, to a Miss Lizzie Jarvis, the inventor said:

"It occurred to me that if I could invent a machine—a gun—which could by its rapidity of fire enable one man to do as much battle duty as a hundred, that it would, to a great extent, supersede the necessity of large armies, and consequently exposure to battle would be greatly diminished."

Undated and unidentified newspaper article in the Gatling Family Papers. Courtesy Mr. John Waters Gatling, Asheville, North Carolina.

By John Hix

STRANGE AS IT SEEMS—

BACKFIRE!

RICHARD JORDAN GATLING INVENTED THE WORLD'S FIRST MACHINE GUN IN AN ATTEMPT TO END WAR!

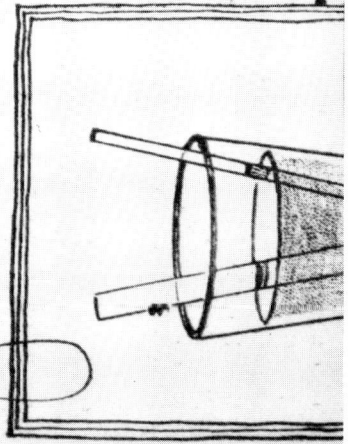

From the September 30, 1936 issue of the St. Petersburg (Florida) Times. Gatling Family Papers, courtesy Mr. John Waters Gatling, Asheville, North Carolina.

111

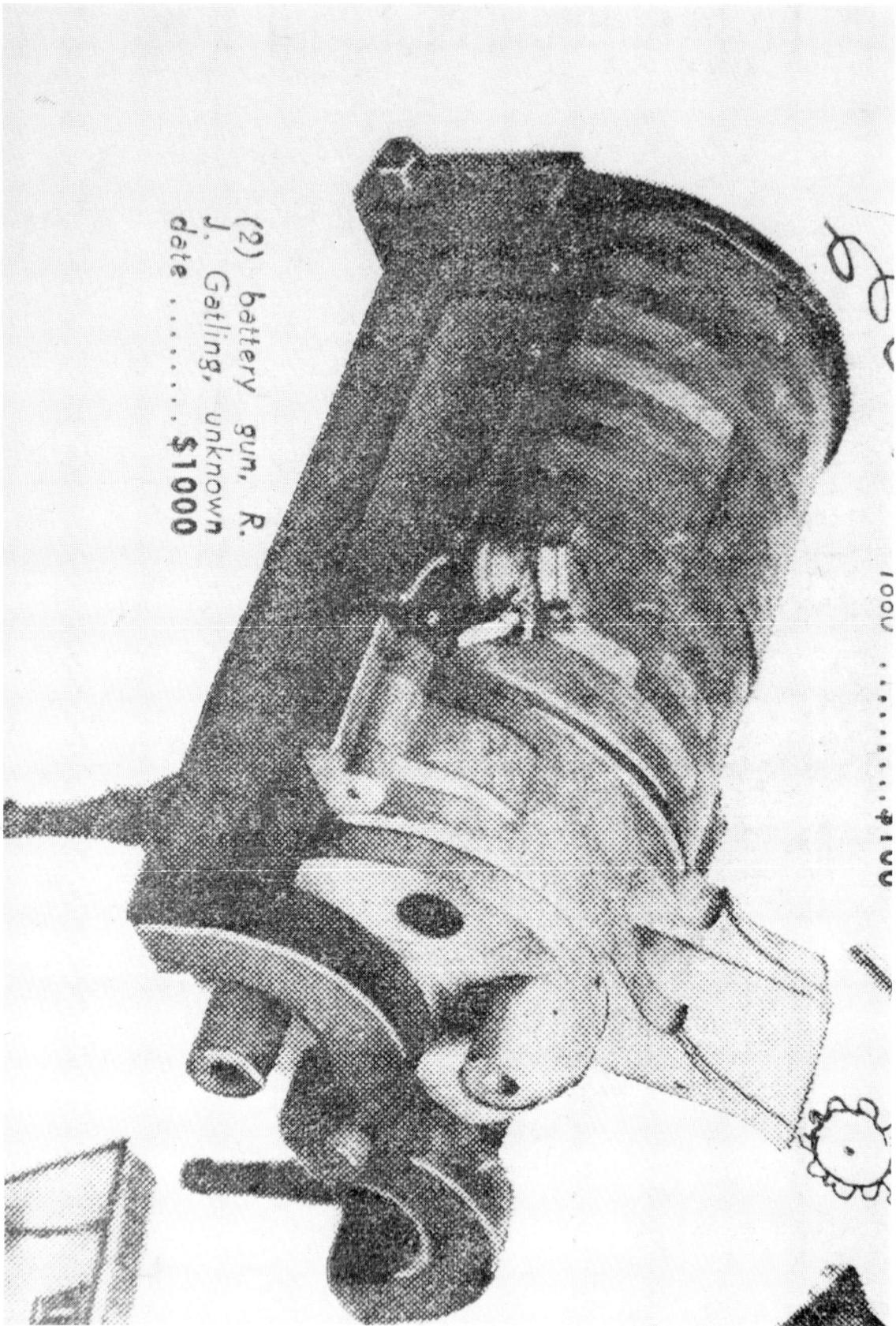

(2) battery gun, R. J. Gatling, unknown date $1000

Photograph of Richard Jordan Gatling's original gun patent model in a Gimbel's department store ad in the New York Herald Tribune, June 2, 1950. The ad featured a number of the 150,000 original patent models from 1820-1890 that the Patent Office sold in 1925 to Sir Henry Wellcome who planned to build a museum in New York City to display them. The stock market crash of 1929 ended Sir Henry Wellcome's plans. After going through a series of owners the collection of patent models in 1942 became the property of a group of investors headed by O. Rundle Gilbert. In 1950 Gilbert made arrangements with Gimbel's department store to sell them. Gatling's original gun patent model sold for $1000, the most expensive of all the patent models sold. After the Gimbel's department store sold them, Gatling's original gun patent model was sold by Gimbel's, John Waters Gatling attempted to locate the purchaser but was unsuccessful.

112

Gatling's Gun in Sale Of Old Patent Models

Models of such inventions as Aaron Diffenderfer's patent sausage-stuffer, a combined henhouse and animal trap, the first machine gun and the first inner-spring mattress are among the 5000 items on display today at Gimbels.

The department store is showing a collection of models sent by inventors to the U.S. Patent Office between the years 1820 and 1890, and they'll be on sale Monday.

The 5000 models are only a sampling of some 150,000 in the store, most of them still crated. The rest will be brought out and put on sale as room is made for them. Gimbels is acting as commission agent for a group of owners, headed by O. Rundle Gilbert, auctioneer with offices at 505 Fifth Ave.

Mr. Gilbert has no idea what is in the still unopened cases. Most of them have been sealed up for 50 years or more.

The Patent Office used to require all inventors to supply a working model, but abandoned that rule in 1890 because of lack of space. The collection that had piled up from 1820 to 1890 was sold in 1925, passed through various hands and now has wound up at Gimbels.

Prices, Mr. Gilbert said, will range from $1 to $1000, the latter for the model of the Gatling machine gun.

That animal trap and henhouse is an interesting item. Each hen's compartment has a door with a sign on it, legible when the hen has opened the door. It reads: "I am out. You may have my egg."

From New York Telegram and Sun, June 2, 1950.

From the New York Herald Tribune, February 6, 1958. Gatling Family Papers. Courtesy Mr. John Waters Gatling, Asheville, North Carolina.

Old Gun Pays Off

FORT WORTH, Tex. — A souped-up version of the Gatling machine gun, in moth-balls for almost a century, is now standard equip-

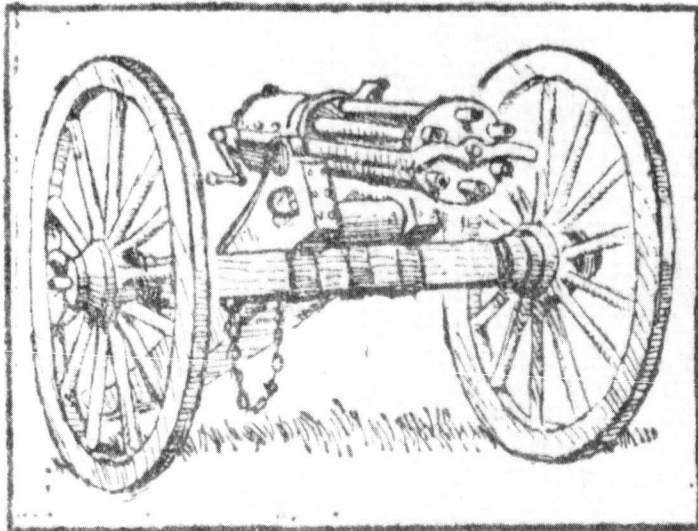

Gatling Gun

ment in the tail of B-58 jet bombers, it was disclosed yesterday. The original gun, hand-cranked, could fire 350 shots a minute. The B-58 version, run by motor, spits out 7,000 rounds a minute. The old Gatling met requirements for a weapon that woudn't melt in the heat of intensive firing.

114

Gatling Gun Goes Jet Age

ABERDEEN, Md. (AP) — A lethal new airborne cannon, patterned on a weapon nearly a century old, was unveiled here yesterday as the answer to the supersonic age's need for rapid firepower.

The 20-millimeter T171, nicknamed "the Vulcan" for ancient Rome's God of Fire, was demonstrated at the Army proving ground by the U. S. Air Force, Army Ordnance, and General Electric, who developed it specifically for use in modern jet aircraft.

The new machine gun incorporates the original design of several rotating barrels in a cluster, patented in 1862 as the Gatling gun, which later revolutionized firepower in the Spanish-American War.

Undated and unidentified newspaper article in the Gatling Family Papers. Courtesy Mr. John Waters Gatling, Asheville, North Carolina.

115

G. E. Steps Up Output of a Modern Gatling Gun for Jets

General Electric engineer inspects the new 30-mm. version of the Air Force's Vulcan gun

Few would consider General Electric Company a munitions maker, yet the giant electrical manufacturer is stepping up its activity in the field.

It is today the sole manufacturer of the Vulcan gun, a 20 and 30 millimeter direct descendant of the famed Gatling gun patented in 1862 by Richard J. Gatling. G. E. received yesterday a $7,116,803 contract to manufacture the gun for jet aircraft at its Burlington, Vt., missile and ordnance systems department. This is in addition to a $17,-000,000 contract awarded last month.

Initial work on the gun was begun in 1946. The 20-mm. version was announced Aug. 28, 1956. The 30-mm. "big brother" boosts the striking power of jet aircraft three times over its predecessor. It was unveiled on July 31 before the Air Force Association at Washington, D. C.

Like the Gatling gun, both new models use a rotating cluster of six barrels, external power sources and an electric drive. Development work on early models has reduced the number of parts from 779 to 448, permitting field stripping and reassembly in thirty minutes. It was designed, primarily to overcome tempera-

ture, speed and altitude barriers that made the standard machine guns obsolete.

While the actual rate of fire has not been revealed, military experts said that either model can fire up to 6,000 rounds a minute. This compares with 650 to 800 rounds a minute for the standard M-3 20-mm. machine gun and 400 to 550 rounds for the 30-caliber air-cooled light machine gun used by the infantry.

The possibility that a ground model is being developed is indicated by the fact that the contracts signed thus far have been "for the U. S. Air Force under contract with the Army Ordnance Corps."

Article from <u>New York Times</u>, August 4, 1957. Gatling Family Papers, courtesy Mr. John Waters Gatling, Asheville, North Carolina.

R. J. GATLING.
MACHINE GUN.

No. 502,185.   Patented July 25, 1893.

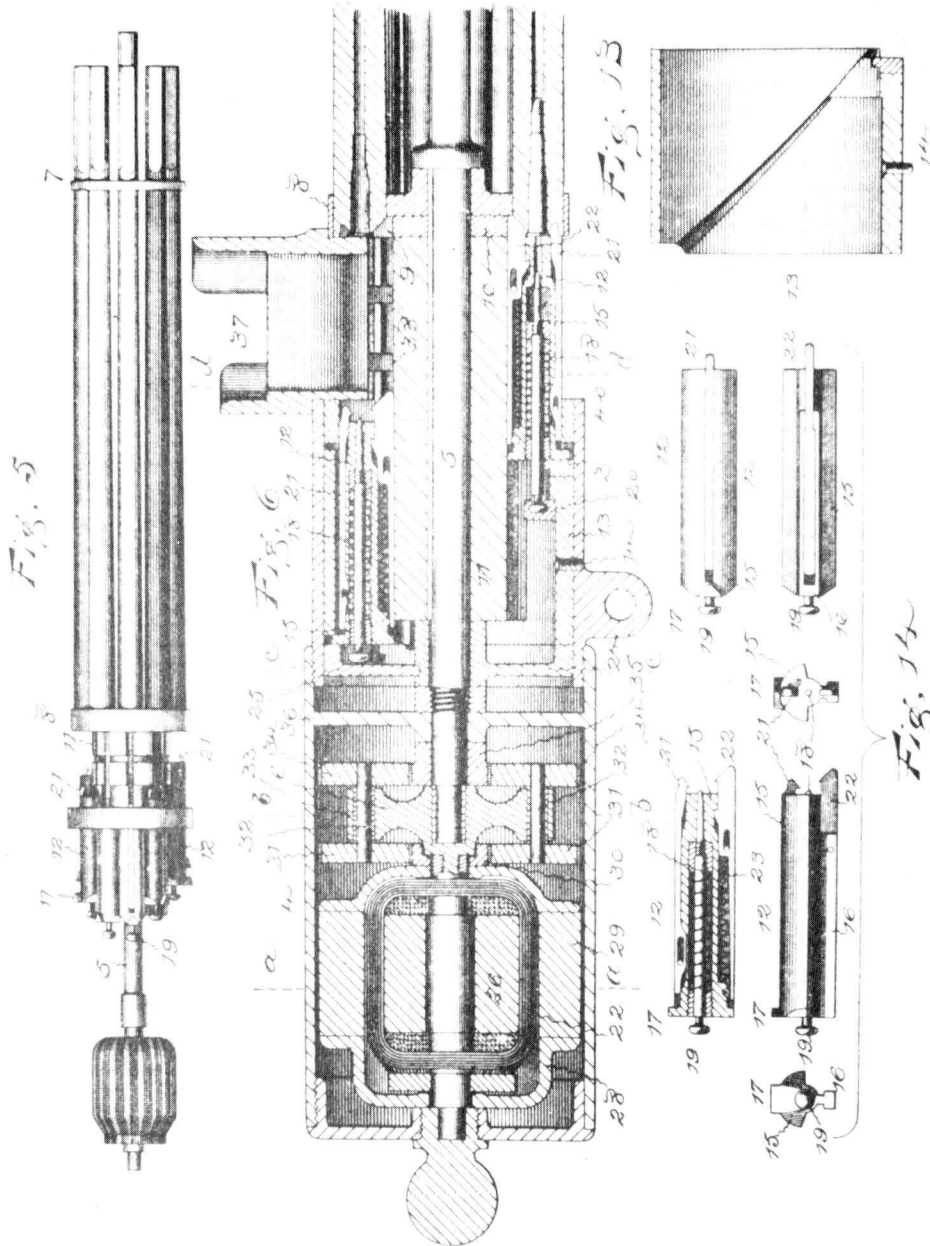

Witnesses: Inventor:
Clarence E. Buckland,   Richard J Gatling by
P. A. Phelps.   Harry R. Williams
                                                      att.

Drawings for Richard Jordan Gatling's electric machine gun, patented July 25, 1893. This version of his gun fired 3,000 rounds a minute and was the basis for the Vulcan gun now in use by United States armed services. Courtesy National Archives, Washington, D.C.

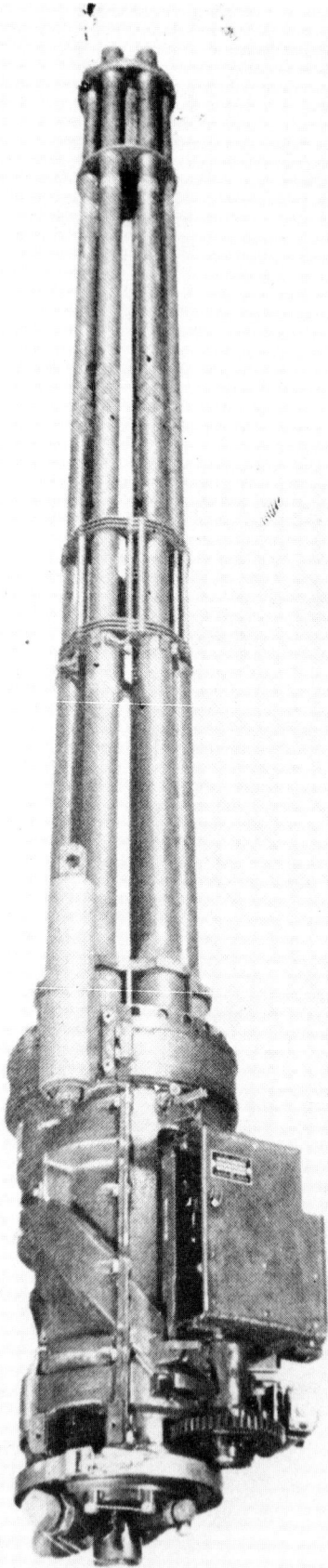

United States Department of Defense photograph of the Vulcan gun.

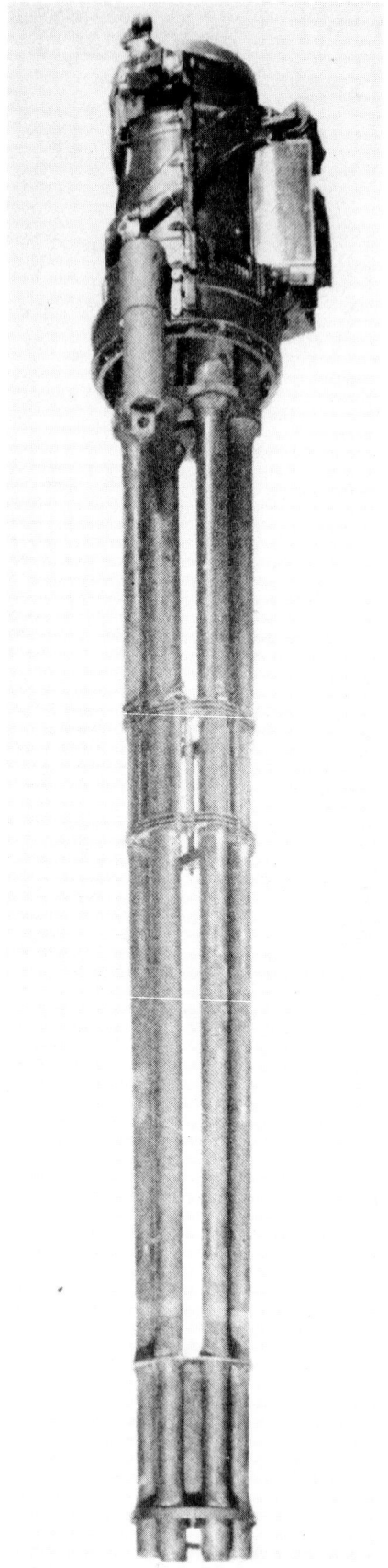

United States Department of Defense photograph of the Vulcan gun.

119

Vulcan (Gatling) gun on US Navy ship, Norfolk, Virginia. The gun fires over 6,000 rounds per minute. Photograph by E. Frank Stephenson, Jr., July 1991.

One of the last photographs of Mrs. Richard Jordan Gatling. She died on September 26, 1908 in New York City and was buried in Crown Hill Cemetery, Indianapolis, Indiana. Courtesy Mr. John Waters Gatling, Asheville, North Carolina.

Lamp given to Mrs. Richard Jordan Gatling by the President of France when Mrs. Gatling accompanied her husband to Europe in 1877. Courtesy The Murfreesboro Historical Association, Murfreesboro, North Carolina. Photograph by E. Frank Stephenson, Jr.

Watch fob belonging to Richard Jordan Gatling. Gatling is wearing the watch fob in the photograph on page 71. Courtesy The Murfreesboro Historical Association, Murfreesboro, North Carolina. Photograph by E. Frank Stephenson, Jr.

Stickpin belonging to Mrs. Richard Jordan Gatling. Courtesy The Murfreesboro Historical Association, Murfreesboro, North Carolina. Photograph by E. Frank Stephenson, Jr.

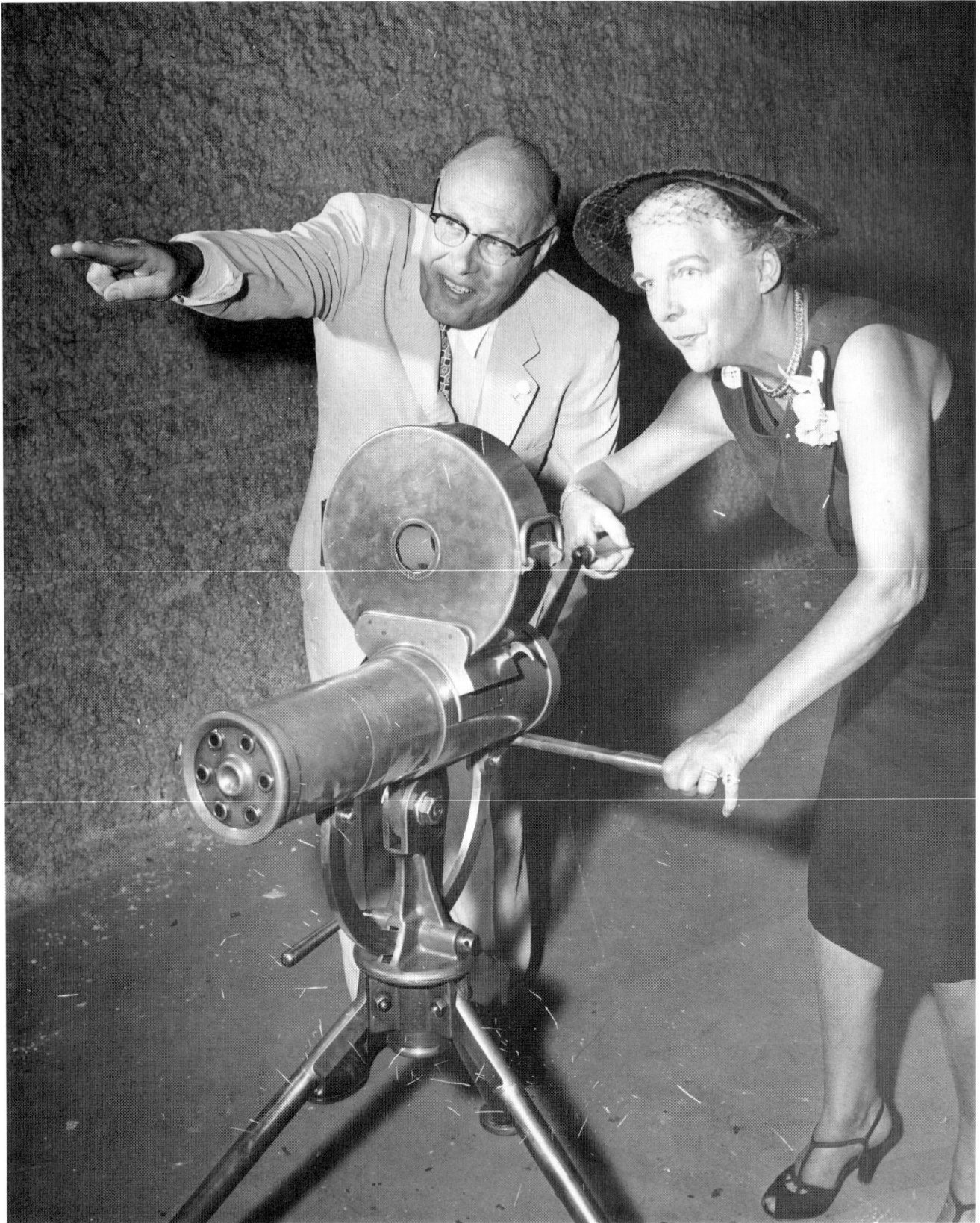

Mr. and Mrs. John Waters Gatling shown here shooting a Gatling gun on August 15, 1959 at the shop of H. H. Harris Guns, State Street, Chicago. Photograph by Ralph Walters, photographer for the Chicago <u>Sun-Times</u>. Courtesy Mr. John Waters Gatling, Asheville, North Carolina.

The Murfreesboro Historical Association's restored William Rea Store housing the Gatling museum and other exhibits on Murfreesboro, North Carolina and the surrounding area. Photograph by E. Frank Stephenson, Jr.

Gatling gun model on display in the Smithsonian Museum, Washington, D.C. Photograph by E. Frank Stephenson, Jr.

The citizens of Como, North Carolina where the Richard Jordan Gatling story began, have not forgotten their most famous son and his invention of the Gatling gun. The art work for the sign was done by Molly Eubank with the woodwork and landscaping by James and Patricia Marshall. Photograph by E. Frank Stephenson, Jr., 1993.

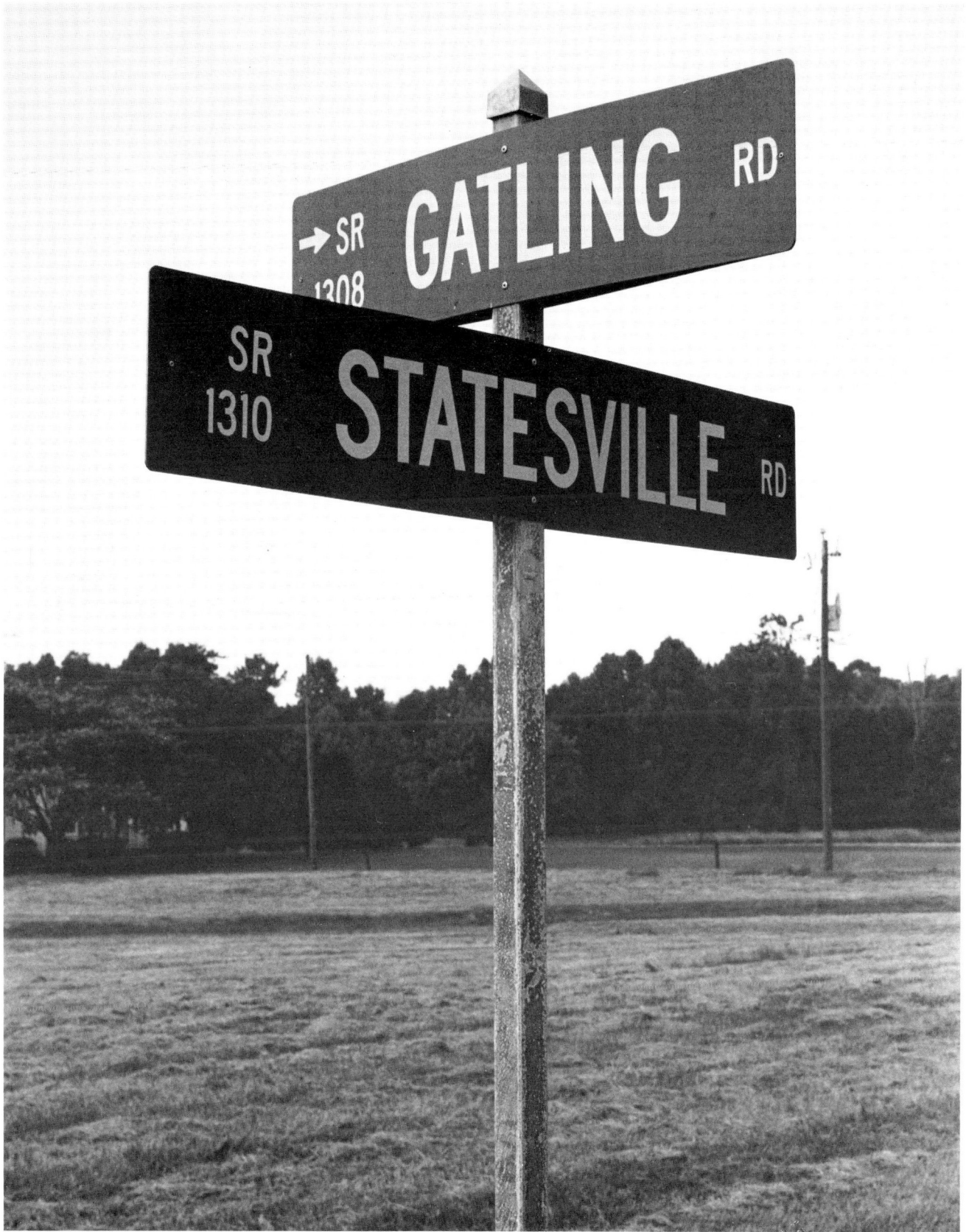

North Carolina state road 1308, now known as Gatling Road. Local people through the years had called the former raceway for the Gatling family's race horses, "Gatling Avenue". Photograph by E. Frank Stephenson, Jr., 1993.

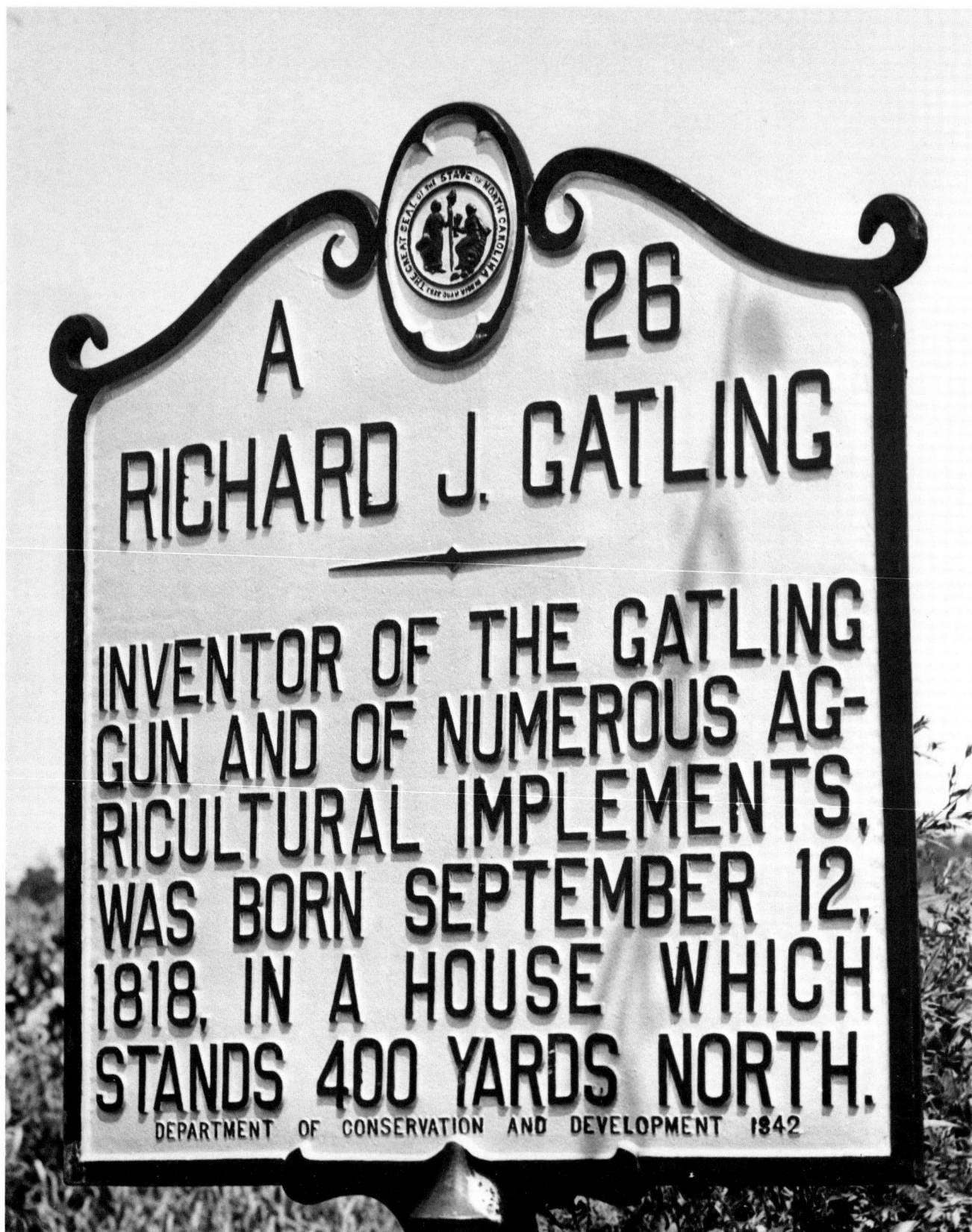

North Carolina historical marker located on US 258 two miles south of Como, North Carolina. At the time of the marker's erection in the 1940s, it was erroneously assumed that the "Great House" was the birthplace of Richard Jordan Gatling. Photograph by E. Frank Stephenson, Jr., 1982.

1987 photograph of a group of Gatling "Cousins" in the restored Gatling room of The Murfreesboro Historical Association's restored William Rea Store. Photograph by Colbert P. Howell, Raleigh, North Carolina.

Steel engraving of Richard Jordan Gatling appearing in <u>Representative Men of The Day</u>, Atlantic Publishing and Engraving Company, New York, 1892. Courtesy Mr. John Waters Gatling, Asheville, North Carolina.